T0318260

Multiple Objective Analytics for Criminal Justice Systems

Emerging Operations Research Methodologies and Applications

Series Editors: Natarajan Gautam, Texas A&M, College Station, USA, A. Ravi Ravindran, The Pennsylvania State University, University Park, USA

Multiple Objective Analytics for Criminal Justice Systems
Edited by Gerald W. Evans

Design and Analysis of Closed-Loop Supply Chain Networks
Edited by Subramanian Pazhani

For more information about this series, please visit: https://www.routledge.com/Green-Engineering-and-Technology-Concepts-and-Applications/book-series/CRCGETCA

Multiple Objective Analytics for Criminal Justice Systems

Gerald W. Evans

CRC Press
Taylor & Francis Group
Boca Raton London New York

CRC Press is an imprint of the
Taylor & Francis Group, an **informa** business

First edition published 2021
by CRC Press
6000 Broken Sound Parkway NW, Suite 300, Boca Raton, FL 33487-2742

and by CRC Press
2 Park Square, Milton Park, Abingdon, Oxon, OX14 4RN

© 2021 Gerald W. Evans
CRC Press is an imprint of Taylor & Francis Group, LLC

Library of Congress Cataloging-in-Publication Data

Names: Evans, Gerald W., author.
Title: Multiple objective analytics for criminal justice systems / Gerald W. Evans.
Description: Boca Raton : CRC Press, 2021. | Series: Emerging operations research methodologies and applications | Includes bibliographical references and index.
Identifiers: LCCN 2020043162 (print) | LCCN 2020043163 (ebook) | ISBN 9780367517342 (hardback) | ISBN 9781003054993 (ebook)
Subjects: LCSH: Criminal justice, Administration of--Statistical methods. | Multiple criteria decision making.
Classification: LCC HV7415 .E93 2021 (print) | LCC HV7415 (ebook) | DDC 364.072/7--dc23
LC record available at https://lccn.loc.gov/2020043162
LC ebook record available at https://lccn.loc.gov/2020043163

ISBN: 978-0-367-51734-2 (hbk)
ISBN: 978-0-367-51735-9 (pbk)
ISBN: 978-1-003-05499-3 (ebk)

Typeset in Times
by Deanta Global Publishing Services, Chennai, India

In memory of Tom Morin, Professor of Industrial Engineering, Purdue University, 1975–2020.

CONTENTS

PREFACE

Criminal justice systems deal with the investigation and apprehension of the accused; prosecution and pre-trial activities; adjudication and sentencing of guilty criminals; and imprisonment, probation, parole, and restitution for victims and rehabilitation. There is no single criminal justice system in the United States, but many interacting systems.

These systems are complex and are therefore difficult to design and operate. The complexities result from (1) their many interacting parts; (2) their interactions with other systems such as those from social welfare, education, health care, both physical and mental, and others; (3) their many decision makers and stakeholders, which result in the existence of multiple conflicting objectives; (4) their dynamic and probabilistic nature, (5) the fact that changes (through for example new laws) are typically made in a highly partisan political environment; and (6) the fact that good data is often difficult to obtain.

The various methodologies associated with analytics and operations research, especially those methodologies that consider the multiple, conflicting objectives associated with the complex decision problems, allow for the accurate modeling of these systems. These models allow one to make the important decisions required for the design and operation of these systems. Unfortunately, the application of analytics in this area of criminal justice, especially as compared to applications of analytics in other areas, has been limited. As noted by Blumstein (2007), criminal justice is one of the most "primitive social systems" regarding the application of analytic methods.

This book is meant to be used by those who would like (1) an introduction to criminal justice systems and (2) an illustration of how some of the various methodologies of multiple objective analytics can be used to address specific issues in these systems. As such, the book should be of interest to faculty, students, and researchers in schools/departments of criminal justice, law, public affairs, political science, industrial engineering, and management. In addition, the book should be of use to government analysts who study the effects of programs and laws in criminal justice systems.

Due to the brevity of this book, not all areas involving the design and operation of criminal justice systems are covered. Instead, selected programs, policies, and laws are investigated from an analytical standpoint.

The book contains four chapters. Chapter 1 provides an overview of criminal justice systems, including sources of data and endemic problems with criminal justice in the United States. Chapter 2 discusses important decisions in the design and operation of criminal justice systems, and how the various methodologies of analytics can aid in the decision-making process. Included in this chapter are illustrative examples of problem structuring to address (1) juvenile crime and (2) the setting up of a syringe exchange program.

Chapters 3 addresses decision issues associated with the areas of (1) privatization of jails and prisons, and (2) bail reform. Comprehensive reviews of each of these areas are included in the chapter, along with illustrative examples involving (1) the assessment and use of a multiattribute value function to evaluate and rank proposals from private companies to operate a state prison and (2) a multiple objective, decision theoretic model for the ranking of alternatives for pretrial conditions for a defendant. The latter example employs a Monte Carlo simulation model and a multiattribute utility function to consider the various costs, probabilities, risks, and objectives of the situation.

Chapter 4 addresses two federal laws, namely the Violent Crime Control and Law Enforcement Act (VCCLEA) of 1994 (sometimes called the Clinton Crime Bill) and the Criminal Justice Reform Act of 2018 (sometimes called the First Step Act). The motivations and the elements of each of these laws are presented. Two examples are presented in the chapter. First, an illustrative example of a cost–benefit analysis of the Community Oriented Policing Services (COPS) program (a major portion of the VCCLEA) is presented. Second, a hypothetical example involving the optimization of two of the controls for the First Step Act is provided. This example also involves the use of a multiattribute utility function interfaced with a Monte Carlo simulation model, to consider the multiple objectives and the risk associated with this decision problem.

Readers with comments, suggestions, or questions are welcome to direct these to the author.

REFERENCE

1. Blumstein, A. (2007). An OR missionary's visits to the criminal justice system. *Operations Research, 55*, 14–23.

ACKNOWLEDGMENTS

Dr. Natarajan Gautam, Professor of Industrial and Systems Engineering (with joint appointment in Electrical and Computer Engineering), Texas A. and M. University; and Dr. A. Ravi Ravindran, Professor Emeritus/Provost's Emeritus Faculty Teaching Scholar at Penn State University provided thoughtful suggestions during the development of the proposal for this book.

Ms. Erin Harris, Senior Editorial Assistant at Taylor & Francis/ CRC Press contributed beneficial suggestions relating to the technical aspects of the book development.

Finally, Dr. Sophia Barkat; Dr. Taimour El-Cheikh, Senior Data Scientist, Contractor to the Defense Counterintelligence and Security Agency; Dr. Matthew J. Evans, Professor of Political Science, Northwest Arkansas Community College; Judge Steven M. Fleece, Indiana trial court judge, retired; and Dr. Aldo McLean, P.E. Professor of Engineering Management and Technology at the University of Tennessee at Chattanooga provided helpful comments on early drafts of this book.

AUTHOR BIOGRAPHY

Gerald W. Evans is Professor Emeritus in the Department of Industrial Engineering at the University of Louisville (UL). His research and teaching interests lie in the areas of multicriteria decision analysis, simulation modeling and analysis, optimization, logistics, and project management.

His previous positions include Industrial Engineer for the Department of the Army, and Senior Research Engineer for General Motors Research Laboratories. He has also served as an ASEE Faculty Fellow at NASA's Langley Research Center and Kennedy Space Center.

Dr. Evans received his BS in Mathematics, his MS in Industrial Engineering, and his PhD in Industrial Engineering, all from Purdue University.

Dr. Evans has performed funded research obtained from organizations such as the National Science Foundation, the Defense Logistics Agency, NASA, the National Institute for Hometown Security, Louisville Metro Government, General Electric, and United Parcel Services among other organizations.

He has published approximately 100 papers in various journals and conference proceedings. His paper: "An Overview of Techniques for Solving Multiobjective Mathematical Programs," published in 1984 was listed as one of the most cited publications in *Management Science* over the last 50 years (from the 2004 commemorative CD: *Celebrating 50 years of Management Science*, INFORMS Publication). In addition, his co-authored paper: S.J. Ellspermann, G.W. Evans, and M. Basadur, "The Impact of Training on the Formulation of Ill Structured Problems," *Omega* 35:2, pp. 221–236,

April 2007, received the 12th Annual Citation of Excellence Award as one of the top 50 management articles out of 15,000 articles published in 2007 (Emerald Management Reviews, 2008).

Dr. Evans is the author of *Multiple Criteria Decision Analysis for Industrial Engineering: Methodology and Applications*, CRC Press, Taylor and Francis Group, Boca Raton, FL, 2017, and co-editor of the book *Analytics, Operations, and Strategic Decision Making in the Public Sector*, Hershey, PA.: IGI Global, 2019. He has also served as co-editor of the Proceedings of the Winter Simulation Conference in 1993 and 1999, and of a special issue of *Computers and Industrial Engineering* titled "Multi-Criteria Decision Making in Industrial Engineering". He has been a reviewer for approximately 25 different journals, an Associate Editor for *the Institute of Industrial Engineers (IIE) Transactions*, a Director of the Operations Research Division of IIE, and a Vice President of IIE.

Dr. Evans has received the Fellow Award and the Operations Research Division Award from the Institute of Industrial and Systems Engineering, the Moving Spirit Award from INFORMS for his work with the UL INFORMS Student Chapter, the Dean's Award for Outstanding Graduate Teaching; and he was a University of Louisville nominee for Outstanding Faculty of Adult Learners for Kentukiana Metroversity Inc. He is listed in American Men and Women of Science, Who's Who in Engineering, and Who's Who in America.

CRIMINAL JUSTICE SYSTEMS IN THE UNITED STATES

1.1 INTRODUCTION

This chapter provides a definition of crime, as well as a classification of, and the costs for crime (discussed in the next section). This is followed by the third section, which discusses the reasons for crime, and the fourth section, which provides a definition for criminal justice systems, as well as a discussion of laws and processes.

The fifth section discusses data associated with criminal justice in the United States as well as its sources. This section contains a table of crime rates in the United States, starting with the year 1960. A new way of interpreting these crime rates is also suggested.

The sixth section of the chapter provides data which compares criminal justice in the United States to other countries, while the seventh section discusses issues (problems or opportunities) associated with criminal justice in the United States. Finally, the chapter ends with a look forward to the remaining chapters.

1.2 CRIME: DEFINITION, CLASSIFICATION, AND COSTS

What constitutes a crime is something that changes with the law over time. One definition is given by:

> *Crime is the breaking of a law for which the criminal justice or some other governing authority prescribes punishment.* (page 13, Rennison and Dodge, 2018)

One type of classification of crimes is by federal or state. Federal crimes are those that violate laws passed by the Congress; these are tried in federal courts. State crimes are those that violate laws passed by state legislatures, and as such are tried in local courts.

A second type of classification of crimes is by felony, wobbler, misdemeanor, and infraction/violation. Each jurisdiction determines its own subclassifications for felonies and misdemeanors. For example, felonies could be categorized as being of Class A through Class J, from the more to less serious. Misdemeanors could be categorized as being of Class 1 through Class 4, with Class 1 being the most serious and Class 4 being the least serious.

A felony is the most serious type of crime. Conviction of this type of crime will result in a sentence of more than a year in a state or federal prison.

A wobbler is a crime in which the defendant could be found guilty of either a felony or a misdemeanor, depending upon the facts of the case and the elements of the offense. For example, a battery that is impolite, rude, or insulting is a lower-level misdemeanor, but if it causes bodily injury it can become a higher-level misdemeanor; if it causes serious bodily injury it becomes a felony (personal communication with Steven M. Fleece, state of Indiana trial court judge, retired, August 24, 2020).

A misdemeanor is of course less serious than a felony, and conviction on this type of crime will result in a sentence of a year or less in a local jail, suspended on terms of probation. If the terms of the probation are violated, the judge can impose some or all of the suspended sentence.

Finally, an infraction/violation is something like a traffic ticket. It does not result in a jail sentence and is not considered a criminal offense. It can however result in a fine.

The cost to federal, state, and local governments in 2012 for the various functions of criminal justice has been estimated at $280 billion, adjusted to 2016 dollars (GAO Report, 2017). These functions can basically be divided among policing, courts, and jail/prison, including probation services.

These government costs are just a relatively small part of the overall costs of crime. For example, the GAO report referenced above reported that there were four different primary methods for estimating the cost of crime, and that these methods provided the widely varying estimates of $690 billion, $1.57 trillion, and $3.41 trillion as the cost of crime in 2012, again adjusted to 2016 dollars (GAO Report, 2017).

One of the primary reasons for the wide variance in the estimates has to do with how intangible (or nonmonetary) costs are considered. Examples of intangible costs are victim pain, suffering, and lost quality of life; increasing fear in the community; the public's change in behavior as a result of the fear of crime; lost wages from incarceration; and the psychological cost to the family of the incarcerated. DeVuono-powell et al (2015) discuss the cost of incarceration on families.

Regardless of which estimate is closest to the true value, the cost of crime in the United States is massive. Hence, even small improvements in the design and operation of criminal justice systems can result in sizable savings and improvements in quality.

1.3 REASONS FOR CRIME

Many would label this section as **Causes** for Crime. However, when one gives the subject some thought, the word **Reasons** appears more appropriate. A cause is something that leads, with little deviation, to something else, but a reason is a rationale, explanation, or justification for something. For example, one could say that poverty might be a reason for crime, but not a cause for crime. There are many people who grow up in poverty who do not become criminals.

The list of reasons for crime is long, and includes the following: poverty, peer pressure, drug and alcohol addiction, religion, family conditions, the society, unemployment, and deprivation (Top 10 Reasons for Crime, 2019). Other reasons for crime include mental illness, misogyny (in the case of domestic violence), and desire for power, wealth, or opportunity (especially in the case of white-collar crime). Some of these are obviously overlapping in nature. For example, deprivation, unemployment, and poverty would all be strongly associated with each other. Someone who is unemployed may well also be impoverished. Such a person may turn to selling drugs to support themselves financially.

The rationale for being concerned about the reasons for crime is obvious. In general, when these reasons are addressed, not only can crime be reduced, but social conditions in general can be improved. With respect to analysis of the effects of a decision made to improve the criminal justice system, having so many reasons for crime makes this analysis difficult. For example, a portion of the Violent Crime Control and Law Enforcement Act of 1994 resulted in the hiring of approximately 88,000 additional police. Judging the effects on crime as a result of this hiring was difficult because of the reasons mentioned above; for example, how much of the decrease in crime during the late 1990s was a result of improving economic conditions, and how much was a result of the increased number of police? This question is addressed in detail in Chapter 4 of this book.

1.4 CRIMINAL JUSTICE SYSTEMS: DEFINITION, LAWS, RESOURCES, AND PROCESSES

There are many different definitions for "criminal justice systems". These definitions can focus on the functions of the system, the organizations associated with the system, or both. A definition that focuses on the functions is:

> A criminal justice system is *the system of law enforcement that is directly involved in apprehending, prosecuting, defending, sentencing, and punishing those who are suspected or convicted of criminal offenses.* (Criminal Justice System, n.d.)

Since recidivism is a huge problem for criminal justice, the above definition might be thought of as lacking since it does not include **rehabilitation** as an important function of criminal justice systems. In addition it does not include **restitution**, another important function of the criminal justice system.

In the United States there are several criminal justice systems. This corresponds to the fact that there are two categories of court systems in the United States: the federal system and the state system. Each of these types of systems is hierarchical in nature.

The federal system consists of the US Supreme Court, 13 federal courts of appeal, and 94 federal district courts assigned to specific areas of the country (page 182, Rennison and Dodge, 2018). As the name indicates, the federal court system addresses federal crimes.

Each state in the country has its own state court system. These state systems have a state supreme court at the highest level, which has jurisdiction over superior courts, municipal courts, county circuit courts, county courts, and district courts. These latter five types of courts are often just generically called trial courts. There are also intermediate courts, called courts of appeal, which handle most appeals from the trial court decisions. Only an extremely limited number of appeal cases, such as those involving the death penalty, can go directly from a trial court to the state supreme court.

Specialty courts for criminal justice, typically at the local level, also exist; these courts include juvenile courts, drug courts, mental health courts, and veterans' courts. Initially, these courts were started by local trial court judges who saw the need for them. More recently, states have given grants to develop these specialty courts.

Finally, there are also tribal courts, generally having jurisdiction over crimes committed on reservation land.

The three main components of a criminal justice system are (1) police; (2) the courts including defense lawyers and prosecution; and (3) jails, prisons, probation agencies, and parole systems. Overlaying the system are laws, acts, statutes, policies, and procedures which in effect control the inflow and outflow of the system.

Many use the terms law, statute, and act interchangeably. The word, law, has several different definitions. In a generic sense, **law** can be defined as *a binding custom or practice of a community: a rule of conduct or action prescribed or formally recognized as binding or enforced by a controlling authority* (Merriam-Webster Dictionary, n.d.).

An **act** is defined as *A written law, formally ordained or passed by the legislative power of a state, called in England an "act of Parliament," and in the United States an "act of Congress" or of the "legislature"* (Black's Law Dictionary, 2019). An act is something that evolves as it passes through the legislature. The ultimately enacted version of an act becomes a statute.

The resources associated with a criminal justice system include police, judges, jurors, prosecuting attorneys, defense attorneys, jails, prisons, and the equipment and staffs associated with each of the functions.

Figure 1.1 illustrates the various activities and processes associated with the flow of an accused person (and possibly later in the process a convicted person) through the criminal justice system. The sequence of major activities as shown is (1) entry into the system (through commission of a crime and subsequent arrest); (2) pretrial services, adjudication/determination of guilt or innocence; and (3) consequences of conviction (sentencing and sanctions), and corrections. Two notable features of Figure 1.1 are the variety of paths and exit points through the process. In addition, various resources can be identified as important requirements for conducting the various respective activities. Finally, the inflow to the system can, in some sense, be controlled by the sets of federal and state laws/statutes in existence.

1.5 DATA RELATING TO CRIMINAL JUSTICE IN THE UNITED STATES

1.5.1 Data and Models for Decision Making

This book is concerned with making decisions about the design and operation of criminal justice systems in the United States. Models which are representations of the respective systems are used as an aid in this process. To build these models, one must have appropriate data

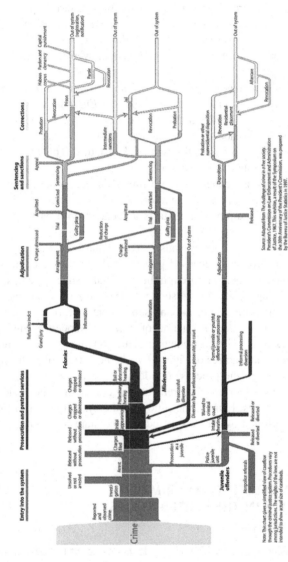

Figure 1.1 Flowchart of process within the criminal justice system (retrieved on April 7, 2020, from the Office of Justice Programs, Bureau of Justice Statistics, at https://www.bjs.gov/content/largechart.cfm).

or information about the relevant system(s). Data can be of one of two types: qualitative or quantitative in nature. Examples of qualitative data would be things like the processes followed when a person is arrested for an alleged crime. Examples of quantitative (or numerical) data would be the percentage of defendants of a socioeconomic class who appear for their court date after being released on bail.

One important aspect of data used for building a model for decision making is categorization of that data. For example, it has been shown that there is a strong relationship between age and crime. Hence, to build a model which will test the effects of a new law related to some type(s) of crime, demographic information related to age and crime is important to track.

In addition to using data for the construction of models, data related to performance can be useful in tracking the effectiveness of a new law/policy.

Data collected and used regarding criminal justice can be categorized according to the various functions shown in Figure 1.1. Following from left to right in Figure 1.1, data can be categorized according to that related to (1) crime and arrests; (2) pretrial and trial services; and (3) jail, prison, probation, and parole. Much of the data collection on criminal justice systems in the United States is accomplished through the Bureau of Justice Statistics of the Department of Justice. As discussed below, data is gathered through the many local agencies throughout the country, and then transmitted to the Bureau of Justice Statistics for compilation at the national level.

The Bureau of Justice Statistics web page on "all data collections" (All Data Collections, n.d.) lists the following categories of relevant information: Corrections, Courts, Crime Type, Criminal Justice Data Improvement Program, Employment and Expenditure, Federal, Indian Country Justice Statistics, Law Enforcement, and Victims. Each of these major categories is divided into subcategories. For example, Corrections is divided into the subcategories of total correctional population, local jail inmates, jail facilities, etc. Finally, within each of these subcategories, there is information on (1) data collection and surveys and (2) publications and projects (arising from the data collection and surveys).

The number of publications and projects arising from each of the subcategories is indeed large. In the following paragraphs, we will discuss just a few of the most important of these, within the areas of (1) crime and arrests, (2) courts, and (3) corrections. The reader is referred to the Bureau of Justice Statistics web page for the actual reports.

1.5.2 The United Crime Reporting Program, the Summary Reporting System, and Crime Rates

First, we will consider crime and arrests. The United Crime Reporting (UCR) Program (more recently called the Summary Reporting System) was started by the FBI in 1930. The Program collects data on the numbers of crime, categorizing them under two broad categories: Violent Crimes and Property Crimes. Approximately 17,000 city, county, state, and federal law enforcement agencies (out of about 18,000 total agencies) participate in the program on a voluntary basis, thereby covering the vast majority of the country (pages 35 and 36 in Rennison ad Dodge, 2018). The violent crimes are divided according to murder, forcible rape, robbery, and aggravated assault. The property crimes are divided according to burglary, larceny-theft, and vehicle-theft. These two crime rates are added to give a total crime rate for each year (United States Population and Rate of Crime per 100,000 People 1960–2018, n.d.). Table 1.1 shows the total crime rate, violent crime rate, property crime rate, and other crime rates (per 100,000 population) in the United States from 1960 to 2018.

One of the things to note from Table 1.1 is that crime rates from each category are added without any weighting. That is, the rates for murder, forcible rape, robbery, and aggravated assault are just added together to give an overall rate for violent crime. Similarly, the rates for burglary, larceny-theft, and vehicle theft are added to give the rate for property crime. Finally, the rates for violent crime and property crime are added to give the total crime rate.

A more accurate picture of crime would be given if the overall rates were weighted by the type of crime. For example, clearly, a murder is more serious than a larceny-theft. One way that these rates could be weighted would be by their costs. The RAND Corporation developed a cost of crime calculator that can be used for this purpose (Cost of Crime Calculator, n.d.). This calculator provides an estimated cost for each type of crime as shown in Table 1.1, with appropriate discounting for the year of the crime. For example, the costs associated with each type of crime for the year 2000 are given by:

TABLE 1.1

Crime Rates (Number of Crimes per 100,000 Population in the United States) from 1960 to 2018

Year	Population	Total Crime Rate	Violent Crime Rate	Property Crime Rate	Murder	Forcible-Rape	Robbery	Aggravated Assault	Burglary	Larceny-Theft	Vehicle-Theft
1960	179,323,175	1,887.2	160.9	1,726.3	5.1	9.6	60.1	86.1	508.6	1,034.7	183.0
1961	182,992,000	1,906.1	158.1	1,747.9	4.8	9.4	58.3	85.7	518.9	1,045.4	183.6
1962	185,771,000	2,019.8	162.3	1,857.5	4.6	9.4	59.7	88.6	535.2	1,124.8	197.4
1963	188,483,000	2,180.3	168.2	2,012.1	4.6	9.4	61.8	92.4	576.4	1,219.1	216.6
1964	191,141,000	2,388.1	190.6	2,197.5	4.9	11.2	68.2	106.2	634.7	1,315.5	247.4
1965	193,526,000	2,449.0	200.2	2,248.8	5.1	12.1	71.7	111.3	662.7	1,329.30	256.8
1966	195,576,000	2,670.8	220.0	2,450.9	5.6	13.2	80.8	120.3	721.0	1,442.9	286.9
1967	197,457,000	2,989.7	253.2	2,736.5	6.2	14.0	102.8	130.2	826.6	1,575.8	334.1
1968	199,399,000	3,370.2	298.4	3,071.8	6.9	15.9	131.8	143.8	932.3	1,746.6	393.0
1969	201,385,000	3,680.0	328.7	3,351.3	7.3	18.5	148.4	154.5	984.1	1,930.9	436.2
1970	203,235,298	3,984.5	363.5	3,621.0	7.9	18.7	172.1	164.8	1,084.9	2,079.3	456.8
1971	206,212,000	4,164.7	396.0	3,768.8	8.6	20.5	188.0	178.8	1,163.5	2,145.5	459.8
1972	208,230,000	3,961.4	401.0	3,560.4	9.0	22.5	180.7	188.8	1,140.8	1,993.6	426.1
1973	209,851,000	4,154.4	417.4	3,737.0	9.4	24.5	183.1	200.5	1,222.5	2,071.9	442.6
1974	211,392,000	4,850.4	461.1	4,389.3	9.8	26.2	209.3	215.8	1,437.7	2,489.5	462.2
1975	213,124,000	5,298.5	487.8	4,810.7	9.6	26.3	220.8	231.1	1,532.1	2,804.8	473.7

(Continued)

TABLE 1.1 (CONTINUED)
Crime Rates (Number of Crimes per 100,000 Population in the United States) from 1960 to 2018

Year	Population	Total Crime Rate	Violent Crime Rate	Property Crime Rate	Murder	Forcible-Rape	Robbery	Aggravated Assault	Burglary	Larceny-Theft	Vehicle-Theft
1976	214,659,000	5,287.3	467.8	4,819.5	8.7	26.6	199.3	233.2	1,448.2	2,921.3	450.0
1977	216,332,000	5,077.6	475.9	4,601.7	8.8	29.4	190.7	247.0	1,419.8	2,729.9	451.9
1978	218,059,000	5,140.4	497.8	4,642.5	9.0	31.0	195.8	262.1	1,434.6	2,747.4	460.5
1979	220,099,000	5,565.5	548.9	5,016.6	9.8	34.7	218.4	286.0	1,511.9	2,999.1	505.6
1980	225,349,264	5,950.0	596.6	5,353.3	10.2	36.8	251.1	298.5	1,684.1	3,167.0	502.2
1981	229,146,000	5,858.2	594.3	5,263.8	9.8	36.0	258.7	289.7	1,649.5	3,139.7	474.7
1982	231,534,000	5,603.7	571.1	5,032.5	9.1	34.0	238.9	289.1	1,488.8	3,084.9	458.9
1983	233,981,000	5,175.0	537.7	4,637.3	8.3	33.7	216.5	279.2	1,337.7	2,869.0	430.8
1984	236,158,000	5,031.3	539.2	4,492.1	7.9	35.7	205.4	290.2	1,263.7	2,791.3	437.1
1985	238,740,000	5,207.1	556.6	4,650.5	8.0	37.1	208.5	302.9	1,287.3	2,901.2	462.0
1986	240,132,887	5,480.4	620.1	4,881.8	8.6	38.1	226.0	347.4	1,349.8	3,022.1	509.8
1987	242,288,918	5,550.0	609.7	4,940.3	8.3	37.4	212.7	351.3	1,329.60	3,081.3	529.5
1988	245,807,000	5,664.2	637.2	5,027.1	8.4	37.6	220.9	370.2	1,309.2	3,134.9	582.9
1989	248,239,000	5,741.0	663.1	5,077.9	8.7	38.1	233.0	383.4	1,276.3	3,171.3	630.4
1990	248,709,873	5,820.3	731.8	5,088.5	9.4	41.2	257.0	424.1	1,235.9	3,194.8	657.8

(Continued)

TABLE 1.1 (CONTINUED)

Crime Rates (Number of Crimes per 100,000 Population in the United States) from 1960 to 2018

Year	Population	Total Crime Rate	Violent Crime Rate	Property Crime Rate	Murder	Forcible-Rape	Robbery	Aggravated Assault	Burglary	Larceny-Theft	Vehicle-Theft
1991	252,177,000	5,897.8	758.1	5,139.7	9.8	42.3	272.7	433.3	1,252.0	3,228.8	658.9
1992	255,082,000	5,660.2	757.5	4,902.7	9.3	42.8	263.6	441.8	1,168.2	3,103.0	631.5
1993	257,908,000	5,484.4	746.8	4,737.7	9.5	41.1	255.9	440.3	1,099.2	3,032.4	606.1
1994	260,341,000	5,373.5	713.6	4,660.0	9.0	39.3	237.7	427.6	1,042.0	3,026.7	591.3
1995	262,755,000	5,274.9	684.5	4,591.3	8.2	37.1	220.9	418.3	987.1	3,043.8	560.4
1996	265,284,000	5,087.6	636.6	4,451.0	7.4	36.3	201.9	390.9	945.0	2,980.3	525.7
1997	267,637,000	4,927.3	611.0	4,316.3	6.8	35.9	186.1	382.1	919.6	2,891.8	505.7
1998	270,296,000	4,615.5	566.4	4,049.1	6.3	34.4	165.2	360.5	862.0	2,728.1	459.0
1999	272,690,813	4,266.5	523	3,743.6	5.7	32.8	150.1	334.3	770.4	2,550.7	422.5
2000	281,421,906	4,124.80	506.5	3,618.3	5.5	32	145	324.0	728.8	2,477.30	412.2
2001	285,317,559	4,162.60	504.5	3,658.1	5.6	31.8	148.5	318.6	741.8	2,485.7	430.5
2002	287,973,924	4,125.00	494.4	3,630.6	5.6	33.1	146.1	309.5	747.0	2,450.7	432.9
2003	290,690,788	4,067.00	475.8	3,591.20	5.7	32.3	142.5	295.4	741.0	2,416.50	433.7
2004	293,656,842	3,977.30	463.2	3,514.1	5.5	32.4	136.7	288.6	730.3	2,362.3	421.5
2005	296,507,061	3,900.50	469.0	3,431.5	5.6	31.8	140.8	290.8	726.9	2,287.8	416.8
2006	299,398,484	3,808.10	473.6	3,334.5	5.7	30.9	149.4	287.5	729.4	2,206.8	398.4

(Continued)

TABLE 1.1 (CONTINUED)

Crime Rates (Number of Crimes per 100,000 Population in the United States) from 1960 to 2018

Year	Population	Total Crime Rate	Violent Crime Rate	Property Crime Rate	Murder	Forcible-Rape	Robbery	Aggravated Assault	Burglary	Larceny-Theft	Vehicle-Theft
2007	301,621,157	3,730.40	466.9	3,263.5	5.6	30	147.6	283.8	722.5	2,177.8	363.3
2008	304,374,846	3,669.00	457.5	3,211.5	5.4	29.7	145.7	276.7	732.1	2,167.0	314.7
2009	307,006,550	3,465.50	431.9	3,036.1	5.0	29.1	133.1	264.7	717.7	2,064.5	259.2
2010	309,330,219	3,350.40	404.5	2,945.90	4.8	27.7	119.3	252.8	701	2,005.80	239.1
2011	311,587,816	3,292.50	387.1	2,905.40	4.7	27	113.9	241.5	701.3	1,974.10	230
2012	313,873,685	3,255.80	387.8	2,868.00	4.7	27.1	113.1	242.8	672.2	1,965.40	230.4
2013	316,497,531	3,112.40	379.1	2,733.30	4.5	25.9	109	229.6	610.4	1,901.60	221.3
2014	318,907,401	2,946.10	372	2,574.10	4.4	26.6	101.3	229.2	537.2	1,821.50	215.4
2015	320,896,618	2,885.10	384.6	2,500.50	4.9	28.4	102.2	238.1	494.7	1,783.60	222.2
2016	323,405,935	2,849.10	397.5	2,451.60	5.4	40.9	102.9	248.3	468.9	1,745.40	237.3
2017	325,147,121	2,757.80	394.9	2,362.90	5.3	41.7	98.6	249.2	429.7	1,695.50	237.7
2018	327,167,434	2,580.10	380.6	2,199.50	5	42.6	86.2	246.8	376	1,594.60	228.9

Murder: $7.3 million, Rape: $185,558, Robbery: $57,300, Aggravated Assault: $74,301, Burglary: $11,154, Larceny-Theft: $1,821, and Motor Vehicle Theft: $7,732.

Admittedly, the actual cost for these crimes would vary greatly from one case to another. However, in terms of obtaining a more accurate view of crime over time, it would seem logical to use these cost values as multipliers for the rates shown in Table 1.1. For example, if one used these cost values in a computation for the violent crime rates, property crime rates, and overall crime rates for the years 1996 through 2001, their values would be (given in units of cost in thousands of dollars per 100,000 people):

Violent crime rates (1996 through 2001): $101369, $95355, $88624, $81136, $78469, $78962.
Property crime rates (1996 through 2001): $20032, $19433, $18131, $16504, $15827, $16129.
Total crime rates (1996 through 2001): $121401, $114788, $106756, $97640, $94297, $95091.

Note that, as would be expected with this approach, the overall crime rates are more heavily weighted by the violent crimes. In addition, the decrease in the total crime rate from 1996 to 2001 is computed as 21.7%, as opposed to the value of 18.1% from Table 1.1 in which figures are not weighted by the cost of crime.

1.5.3 Data on the Solving of Crimes

An FBI report (FBI: UCR, 2017 Crime in the United States, n.d.) provided data on the percentages of various types of crimes that were solved by the police in 2017:

Murder and Nonnegligent Manslaughter:	61.6%,
Rape:	34.5%,
Robbery:	29.7%,
Aggravated Assault:	53.3%,
Burglary:	13.5%,
Larceny-theft:	19.2%,
Motor Vehicle Theft:	13.7%.

In general, one could say that these figures do not indicate particularly good results; however, at least the higher percentages are associated with what are considered the most important crimes (i.e., the violent crimes as indicated in Table 1.1: murder and nonnegligent manslaughter, rape, robbery, and aggravated assault). Lopez (2018) notes that there is much variation in the solved percentage for murder and nonnegligent homicide according to locality and race. He notes that black victims accounted for the majority of homicides in 52 of the largest cities in the United States over the previous decade, and that they were the least likely of any racial group to have their murders result in an arrest. In particular, 63% of white victims had their homicide cases solved, as compared to only 47% of black victims in the 52-city study.

One possible reason for poor performance in solving cases is the way police are utilized—that is with more of an emphasis in preventing crimes rather than solving crimes (Lopez, 2018). This issue of how police are utilized is discussed in more detail in Chapter 4, in the section on the Violent Crime Control and Law Enforcement Act of 1994.

1.5.4 The National Incidence-Based Reporting System, the Law Enforcement Officers Killed and Assaulted Program, and the Hate Crime Statistics Program

As noted in 1.5.2, the Summary Reporting System only gives counts of the number of crimes of each type by jurisdiction. Hence, in recent years more detailed reporting systems have been developed, providing details about the crimes committed. These detailed reporting systems, produced on a semiannual basis, are of three types, namely: (1) reports from the National Incident-Based Reporting System (NIBRS), (2) reports from the Law Enforcement Officers Killed and Assaulted (LEOKA) Program, and (3) reports from the Hate Crime Statistics Program.

The NIBRS evolved from the UCR Program by allowing for greater detail in the data collected for each incident of crime. By allowing for this greater detail, the NIBRS provides for the association of the attributes of the crime, victim information, and offender information with a single offense. The NIBRS groups offenses into two categories: Group A (containing 23 different crimes covering 52 offenses) and Group B (covering 10 offenses). An example of a Group A crime is fraud which would cover the offenses of ATM fraud, impersonation, wire fraud, identity theft, and hacking/computer invasion (Page 41, Renison and Dodge, 2018). In 2018, only 44% of the agencies that took part in the UCR Program participated in the NIBRS Program; however, thousands of additional agencies have made commitments to participate by 2021.

The expressed purpose of the LEOKA Program is to collect data and then thoroughly analyze this data to provide training to law enforcement to keep the relevant personnel safe. The focus is on **why** an incident (in which a law enforcement official was injured or killed) occurred so that an appropriate analysis is accomplished. This analysis is supposed to result in training for law enforcement to mitigate the problem (Law Enforcement Officers Killed and Assaulted, 2019).

In 1990, the US Congress passed the Hate Crime Statistics Act, which resulted in the Hate Crime Statistics Program. This program, under the UCR umbrella, provided for the collection of hate crime data under several different bias motivation categories, including race/ethnicity/ancestry bias, religious bias, sexual orientation bias, gender identity bias, disability bias, and gender bias. Data collected in 2018 indicated that approximately 60% of the hate crimes committed in 2018 fell into the category of race/ethnicity/ancestry bias (Oudekerk, 2019).

Another category of UCR Program reporting is the Supplementary Homicides Reports (SHR). As the name indicates, these reports provide specific data on homicides committed in the United States. In addition to the "numbers data" provided by the UCR reports, the SHR system provides data on specific characteristics of both the offender and the victim, such as age, sex, race, ethnicity along with the relationship between the two; the type of weapon used; and the circumstances surrounding the incident (The Nation's Two Measures of Homicide, 2014).

The above programs and reports have as their source the law enforcement agencies of the United States. The National Crime Victimization Survey (NCVS) on the other hand surveys 49,000–77,400 individual households twice per year. The survey is administered by the Bureau of Justice Statistics and is conducted by the US Census Bureau. The survey involves the collection of information about nonfatal personal crimes and household property crimes. In addition to information about the characteristics of the crime, the survey collects data on the victim, the offender, and whether the crime was reported to the police (Data Collection: National Crime Victimization Survey (NCVS), 2018).

In addition to publishing data and reports on crime, the Bureau of Justice Statistics publishes data/reports on court systems. The web page listed under courts enumerates the following reports/programs/surveys/censuses: Census of Problem-Solving Courts, Census of Public Defender Offices, Census of State Court Organization, Civil Justice Survey of State Courts, Court Statistics Project, Juveniles in Criminal Court, National Judicial Reporting Program, National Survey of Indigent Defense Systems, National Survey of Prosecutors, and State Court Processing Statistics (Bureau of Justice Statistics: Courts, n.d.).

The Census of Problem-Solving Courts addresses the operation of the so-called specialty courts. These courts are defined as those which "use therapeutic justice to reduce recidivism" (Data Collection: Census of Problem-Solving Courts, 2012). An example of such a court would be a drug court. The Census of Problem-Solving Courts provides data on the number of participants, services provided, and benefits associated with the court.

The Court Statistics Project provided annual data from 1975 to 2007 on state courts, such as structure, jurisdiction, caseload volume, and trends. One of the advantages of this project is that, since the data is translated into a common framework, it allows for the comparison of the various states with respect to the operation of their state courts.

Reports published by the Bureau of Justice Statistics related to corrections include Jail Inmates in 2018 (Zeng, 2018), and the 2018 Update on Prisoner Recidivism: A 9-Year Follow-Up Period (2005–2014) (Alper, Durose, and Markman, 2018).

Some interesting facts from the report, Jail Inmates in 2018, include the following: the number of jail inmates at midyear 2018 (738,400 inmates) declined by 12% from 2008 (785,500 inmates) and the jail population in 2018 was 50% white, 33% black, and 15% Hispanic. While the percentage of the jail population that was black in 2018 was still disproportionately high compared to the percentage in the entire population (at 12.3%), this still represents a significant reduction from 45.8% of the jail population from 2008 (a 28% reduction from 45.8% to 33%).

In addition to the large number of people in jail or prison, there is typically even a larger number on probation or parole in the United States. For example, in 2014, there were 3,864,100 people on probation and 856,900 parolees (Kaeble et al, 2015).

The statistics regarding recidivism are complex in nature because of the different ways that recidivism can be defined, among other things. For example, one can address recidivism among those released from prison/jail, or among those on probation. One can also classify recidivists as (1) those who are rearrested for a technical violation, (2) those who are rearrested for a new crime, or (3) those who are rearrested and returned to prison.

Examples of technical violations for those on probation include failure to report, moving without permission, and failing a blood/urine test. Something like failing to report is sometimes just a problem with transportation. Many reformists have suggested some relaxation of the restrictions resulting from these technical violations.

Some of the notable information obtained from the Update on Prisoner Recidivism: A 9-Year Follow-Up Period includes the fact that 44% of the released prisoners from the study were re-arrested during the first year following release. This fact provides a clear indication of the need for greatly enhanced rehabilitation programs for prisoners, both from the standpoint of benefitting the convicted and the public.

Sipes (2018) provides an excellent summary of many of the reports on recidivism, Examples of some of the data from his summary report include the following: 68% of released prisoners were re-arrested for a new crime within three years of release from prison, and 77% were re-arrested within five years. Of course, not everyone who is re-arrested is convicted, but 49.7% of inmates released were returned to prison within three years because of either a technical violation or conviction of a new crime, while 55.1% were returned to prison within five years. These figures point out the importance of rehabilitation.

While there is much useful information on the wide variety of reports that are available from the Bureau of Justice Statistics, there are obvious difficulties with some of the reports. For example, much of the data obtained to provide some of the reports (e.g., the NIBRS reports) was self-reported, and obtained on a voluntary basis; of course, given the large number of agencies (with already heavy workloads) from which data was obtained, such an approach was almost a necessity. One would expect however that agencies with good performances would more likely provide the relevant data.

Second, for many reports, the numbers provided would only be a small part of the overall story. For example, follow-ups on the prisoner recidivism reports would involve extensive questioning of both recidivists and non-recidivists to determine **why** recidivism occurs in certain cases, and **why not** in other cases. Results from such a process would be useful in the design of rehabilitation programs. While there certainly are reports/papers in existence which involve the questioning of former prisoners in this regard, it appears that more effort along these lines would be extremely useful.

Finally, it appears that the generation of reports which involved the use of longitudinal data would be extremely useful. Such an approach was suggested by Blumstein and Cohen (1987) a few decades ago in terms of predicting criminal behavior. Most of the reports discussed above address one area of the criminal justice system (crime and arrests, courts, or corrections). It appears that data collected from individuals as they proceed through the entire system would be useful in establishing appropriate correlations.

Of course, there are trade-offs involved in addressing the three difficulties mentioned above—useful reports generated from addressing these issues would require much additional effort and increased staffing levels.

1.6 DATA COMPARING CRIMINAL JUSTICE IN THE UNITED STATES TO OTHER COUNTRIES

Criminal justice varies from one country to another based on many factors, including history, governmental structure and leadership, demographics, resources, etc. Comparing different countries in this area is a difficult task for many reasons, not the least of which is the large number of performance measures involved.

One comprehensive source that compares various countries with respect to criminal justice is the World Justice Project Rule of Law Index 2020 (World Justice Project Rule of Law Index 2020, 2020). The Index provides normalized scores and a ranking of 128 countries according to their respective performances with regard to the rule of law. The scoring and subsequent ranking are based upon four principles: accountability, just laws, open government, and accessible and impartial dispute resolution, which are then translated into eight factors: constraints on government powers, absence of corruption, open government, fundamental rights, order and security, regulatory enforcement, civil justice, and criminal justice. These 8 factors are respectively subdivided into 44 sub-factors.

To obtain raw data for the sub-factor scoring, surveys were taken of 4,000 legal practitioners and experts and of 130,000 households. Basically, each of the 44 sub-factors can be considered qualitative in nature. For example, under the eighth factor (criminal justice), the seven sub-factors are: criminal investigation system is effective, criminal adjudication system is timely and effective, correctional system is effective in reducing criminal behavior, criminal justice system is impartial, criminal justice system is free of corruption, criminal justice system is free of improper government influence, and due process of the law and rights of the accused. One thing to note about this list of seven sub-factors is that none of them have a direct relationship with a measure of fairness for the victim(s) of a crime. On the other hand, the important problem of recidivism is considered through the third sub-factor, *correctional system is effective in reducing criminal behavior.*

An example of a quantitative measure for "criminal adjudication system is timely and effective" would be something like "average amount of time from arraignment to start of trial". Of course, one can see the value in the use of these qualitative sub-factors in the facts that (1) a qualitative sub-factor can represent many quantitative measures simultaneously and (2) the responder to a survey would not necessarily need a lot of detailed knowledge about the relevant criminal justice system in order to score one of these qualitative factors, as opposed to the detailed knowledge needed to evaluate several detailed quantitative factors.

Examples of the scores and rankings for selected countries are shown in Table 1.2. Of note in the rankings is that the United States is ranked 21st, having fallen out of the top twenty (at number 20) from the 2019 ranking. China is ranked 88th and the Russian Federation is ranked 94th.

In addition to an overall ranking, the World Justice Project provides rankings with respect to each of the eight factors. For example, the United States ranks 22nd on constraints on government power, 19th on absence of corruption, 13th on open government, 26th on fundamental rights, 28th on order and security, 20th on regulatory enforcement, 36th on civil justice, and 22nd on criminal justice.

Finally, the Index provides rankings for countries grouped according to specific characteristics such as income. As would be expected, the average ranking for high-income countries, such as the United States, is much higher than the countries grouped in each of the other income levels.

In addition to the fact that the sub-factors are qualitative in nature (almost out of necessity as noted above), the way that the factor scores are combined to provide an overall score is somewhat arbitrary in nature. An alternative approach which used something like

TABLE 1.2
Scores and Ranking of Selected Countries from
the World Justice Project Rule of Law Index 2020

Country	Normalized Score	Ranking (out of 128 countries)
Denmark	.9	1
Norway	.89	2
Germany	.84	6
Canada	.81	9
United Kingdom	.79	13
Japan	.78	15
France	.73	20
United States	.72	21
Italy	.66	27
Poland	.66	28
Greece	.61	40
South Africa	.59	45
Argentina	.58	48
Panama	.52	63
Brazil	.52	67
India	.51	69
China	.48	88
Russia Federation	.47	94
Venezuela	.27	128

the analytic hierarchy process (pp. 117–135 of Evans, 2017) may be appropriate for combining the factors in an overall score.

Numerical data comparing the United States to other countries in jail/prison systems is provided by the World Prison Brief, Institute for Crime & Justice Policy (Highest to Lowest-Prison Population Total, n.d.). For example, among other information elements, this organization provides data by country, for slightly more than 200 countries on (1) the number of incarcerated individuals per 100,000 population, (2) the number of pre-trial detainees/remand prisoners as a percentage of all prisoners, and (3) the occupancy levels based on official capacity, among other attributes. This data for some selected countries is shown in Table 1.3.

Note that the countries are ranked starting with the largest numerical value ranked first. So, for example, having the United States

TABLE 1.3

Numerical Data on Jails and Prisons for Selected Countries

Country	No. Incarcerated per 100,000 (Rank Out of 223 Countries)	Percent of Prisoners Who Are Pretrial Detainees (Rank Out of 217 Countries)	Prison Occupancy Level (Rank Out of 205 Countries)
Denmark	71 (176)	38.2 (71)	102.7 (114)
Norway	60 (190)	25.2 (128)	87.7 (149)
Germany	77 (171)	20.4 (148)	87.5 (150)
Canada	107 (141)	38.7 (70)	102.2 (115)
United Kingdom	139 (111)	11.7 (197)	111.3 (98)
Japan	39 (206)	11.3 (200)	56.6 (197)
France	104 (147)	29.8 (106)	115.7 (90)
United States	655 (1)	22.5 (141)	103.9 (111)
Italy	102 (150)	31. (101)	120.2 (85)
Poland	199 (75)	11.4 (198)	92.9 (137)
Greece	104 (147)	26.6 (124)	111.2 (99)
South Africa	275 (41)	29.3 (110)	137.4 (64)
Argentina	230 (60)	45.9 (52)	122.1 (82)
Panama	419 (13)	40.7 (64)	122.2 (81)
Brazil	366 (20)	32.8 (94)	167.7 (38)
India	34 (212)	69.4 (16)	117.6 (87)
China	120 (132)	NA	NA
Russia Federation	359 (22)	18.8 (159)	72.8 (180)
Venezuela	178 (89)	63. (21)	153.9 (46)

ranked first among all countries in terms of the number of incarcerated people per 100,000 population is not necessarily a good thing. As seen from Table 1.3, the United States ranks 1st in the number of people incarcerated per 100,000 population, 141st in terms of the percent of prisoners who are pretrial detainees, and 111th in terms of prison occupancy level, which is given as a percentage.

One thing to note about the percent of prisoners who are pretrial detainees is that the total population of prisoners considered are those prisoners who are in local jails, state prisons, and federal prisons, not just local jails. These pretrial detainees are all basically located in local jails and include those who have been accused of a crime but not convicted; this group can be further subdivided into those who have been remanded to jail and those who have been offered bail, but cannot afford it. As reported in an article published in 2016, more than 60% of people in jails in the United States have not been convicted of any crime, and about 90% of these inmates are there because they cannot afford bail (Burdeen, 2016).

1.7 ISSUES ASSOCIATED WITH CRIMINAL JUSTICE SYSTEMS IN THE UNITED STATES

Any large, complex system will have issues and difficulties in terms of its design and operation. In many cases, these issues might be thought of as relative in nature; for example, as compared to criminal justice systems in countries such as Venezuela, Turkey, and Afghanistan (among many others), the criminal justice systems in the United States perform well indeed, at least according to the World Justice Project Rule of Law Index discussed earlier. (Of course, many of these lower-ranked countries have causative issues that are not relevant to the United States, such as unstable governments.) So, what we will discuss in this section might be better termed as opportunities for improvement rather than issues or problems.

Problems with criminal justice in the United States might be described in terms of performance measure values. More specifically, four important issues associated with criminal justice systems in the United States are (1) the costly nature of the system, (2) the inefficient operation of the system, (3) the overcrowded nature of the system, and (4) the unfair nature of the system.

First, as was noted earlier in this chapter, the cost for the various functions of criminal justice in the United States was estimated at $280 billion in 2012, adjusted to 2016 dollars (GAO Report, 2017). The cost of crime itself, in the United States, which the criminal justice system is supposed to reduce, is estimated at one of three different values depending upon the methodology used for estimation: $690 billion, $1.57 trillion, or $3.41 trillion in 2016 dollars (GAO Report, 2017). Given such large amounts of money being spent, even a small fraction of savings would be greatly beneficial.

Second, inefficiencies in the criminal justice system exist in many situations. The most apparent is in the high rate of recidivism associated with released prisoners—as noted earlier, in one study 44% of released prisoners committed another crime within the first year of release. Another example of inefficiency can be seen in the way that jury trials are conducted; juries may be told to arrive for the next day's proceedings at 9 am, only to spend the time from 9 am to 1 pm waiting while the attorneys and judge are working out some issue concerning the trial. Some planning could have saved this time. It is clear why a situation like this would occur—since the judge and attorneys are not paying for the jury's time, there is no incentive for them to value this time.

Third, criminal justice systems in the United States, especially the court systems, jails, and prisons, are very congested. As was mentioned earlier, the occupancy level of jails and prisons in the United States has been measured at 103.9% (Highest to Population Total, n.d.).

Finally, criminal justice systems in the United States may be unfair to accused/convicted persons of specific socioeconomic status, race, and/or nationality. For example, whereas the percentage of blacks in the entire population is only 13%, the percentage of the incarcerated population that is black is 33%. (The gap between the percentage of black prisoners and the percentage of white prisoners, which may be related more to socioeconomic status than race, has decreased in recent years.) However, these percentages do not necessarily mean that the system is unfair. One characteristic of criminal justice systems in the United States is the large number of checks and balances in the system. These checks and balances would provide an argument against the system's unfairness.

1.8 SUMMARY AND A LOOK FORWARD

In this first chapter we have provided definitions for crime and criminal justice systems and estimates of costs associated with each in the United States. We have also identified important data and sources of data about criminal justice systems for the United States and for the United States in relation to other countries. Finally, we presented the five characteristics of criminal justice systems which make them difficult to design and operate.

Chapter 2 comprises three parts. The first part explains some of the methodologies associated with multiple objective analytics. The focus will be on those methodologies that allow for representations of (1) the dynamic and probabilistic nature of criminal justice systems and (2) the multiple objective aspects of system design and operation. The second part of the chapter presents an overview of the important decisions involved in the design and operation of criminal justice systems, especially those to which operations research can contribute. The third part of the chapter provides an overview of past applications of operations research to criminal justice systems.

Chapter 3 presents analyses of specific issues associated with criminal justice systems that have been in the news recently, namely privatization of prisons and bail reform. An overview of these issues and their respective importance is first presented, followed by illustrative applications of analytic methodologies which focus on the aspects of dynamics, uncertainty/risk, and the multiple objective nature of the problem.

Finally, Chapter 4 provides a retrospective discussion and analyses of two important laws that have been enacted over the last few decades: The Violent Crime and Law Enforcement Act of 1994 (Clinton Crime Bill) and its Three-Strikes Provision, and the Criminal Justice Reform Act of 2018 (First Step Act). Again, the focus is on the important aspects of dynamics, uncertainty/risk, and multiple objectives.

ANALYTICS AND CRIMINAL JUSTICE SYSTEMS

2.1 INTRODUCTION

Chapter 2 provides a discussion of analytics, especially those methodologies involving multiple objective analyses, as it relates to criminal justice systems.

The second section of the chapter describes the various decisions relating to the design and operation of criminal justice systems. A major part of this section is a discussion of the complexities of criminal justice systems which make these decisions difficult.

The third section of the chapter discusses five categories of methodologies for dealing with criminal justice systems: (1) problem structuring methods; (2) methods for generating the multiple objectives and performance measures for a decision problem; (3) cost–benefit analyses; (4) multiattribute value functions and multiattribute utility functions, which allow one to trade-off among multiple performance measures; and (5) simulation modeling and analysis. Illustrative examples of problem structuring for juvenile crime and the determination of performance measures for the design of a syringe exchange program are given.

Finally, the fourth section of the chapter provides a brief overview of applications of analytics (or operations research), as described in the literature, to criminal justice systems.

2.2 DECISIONS ASSOCIATED WITH CRIMINAL JUSTICE SYSTEMS AND WHY THEY ARE DIFFICULT TO MAKE

2.2.1 Decisions in General

Making and implementing good decisions is how a system is improved. A decision can be defined as the selection of a single alternative from among two or more alternatives. A good decision is defined as one which has a deliberative process, is evidence-based, balances the interests of those involved, and is made by the proper authority.

An example of a decision situation with only two alternatives is the situation where a state legislature can either pass a specific law or not pass that law.

An example of a decision situation involving many alternatives would be the situation where a law could have any of several different parameter settings. The First Step Act, which was signed into law in December of 2018 (Criminal Justice Reform, n.d.) is discussed in Chapter 4; this law allocated $37 million for drug treatment programs for prisoners during 2021. The funding of $37 million is a somewhat arbitrary amount—the amount could have been, and given the need, probably should have been, much larger. This amount of funding for drug treatment as part of the First Step Act is just one of many different parameter values for the law, and thus represents a decision that was made from among many different alternative levels of funding.

Decisions are often connected to other decisions, in many cases through their sequencing. Referring to the First Step Act again, decisions were made as to various provisions to be included in the act, the amount of funding to be attached to these provisions, and, finally, as to whether the act was passed by the Congress and signed into law by the President.

2.2.2 Decision Makers and Stakeholders for Criminal Justice Systems

A **decision maker** is someone (or an organization) who has responsibility for making decisions. Such responsibility will usually involve making trade-offs among the various conflicting objectives as part of the decision-making process. For example, for a decision about a new state law, the decision maker would be the state legislature and governor. A judge would be the decision maker in many of the decisions made in a criminal trial. Police are decision makers when an arrest is made or not made.

The **stakeholders** for a criminal justice system are basically those who have a stake in the system's operation. It is important to know who these people are in order to establish performance measures for the system—basically the performance measures will be derived from the values of the stakeholders. To evaluate various alternatives with respect to the design and operation of a criminal justice system, we need to compare the values for the performance measures between the different alternatives.

The stakeholders for a criminal justice system are basically those with a stake in the relevant decision situation. These include, over all decisions for criminal justice systems, virtually all adults (and even some children and teenagers) since the systems are supported by taxpayers. But in order to be more specific the list of stakeholders would include police, police unions, victims, defendants, families of the defendants, the convicted, families of the convicted, judges, jurors, prosecuting attorneys, defense attorneys, state legislators, congressional representatives, judicial staff, prison and jail staff, prison unions, and taxpayers, among others.

For any decision situation, it is important to know who these stakeholders are in order to establish performance measures, as derived from the values of the stakeholders. To evaluate various alternatives with respect to a decision situation for a criminal justice system, we need to compare the values for the objectives and corresponding performance measures between the different alternatives.

For example, consider a decision situation where a judge decides about pretrial conditions (e.g., remand or release on bail) for a defendant. The primary stakeholders would include the defendant as well as the defendant's family, the victim(s) of the crime, and the public in general, among others. The objectives for this decision could include: optimize public safety, maximize days of freedom for the defendant until a verdict is given in the case (remember that the defendant may be innocent), minimize incarceration costs to the government, minimize risk to the victim of the crime, and so on. Associated with these objectives would be the uncertainty related to the guilt/innocence of the accused and with the behavior of the accused prior to trial. Decisions which would be good for one stakeholder might not be good for another, and therefore appropriate methods must be employed to consider the trade-offs between the objectives.

The stakeholders listed in the previous paragraph can be grouped into two categories which we will call the government and the defendant. The government, often termed "the people" in a criminal trial, consists of all the stakeholders listed in the previous paragraph except for defendants and their families. The stakeholder which we call defendant consists of the defendant and their family.

2.2.3 Decisions for Criminal Justice Systems

As with almost all complex systems, such as those associated with supply chains, health care, and manufacturing, decisions pertaining to the design and operation of criminal justice systems in the United States can be categorized along at least two dimensions: locality affected and time frame. Locality affected can be, from smallest to largest: local, state, and national. Time frame can be, from shortest to longest: tactical (up to a few days), operational (from a few days to a few months), and strategic (from a few months to several years). (The time durations given are somewhat flexible in nature.) Decisions made on the national/state level can place constraints on the decisions made on the local/case level. In a similar fashion, decisions made on a strategic time frame can place constraints on decisions made at the operational/tactical time frame. Hence, strategic decisions made on the national level represent the most important decisions for criminal justice.

At the tactical level, a document published by the National Institute of Corrections enumerates several key decision points in the criminal justice system (page 23, A Framework for Evidence-Based Decision Making in State and Local Criminal Justice Systems, 2017). Some of these are given as follows, along with the alternatives in parentheses:

1) Arrest decisions (cite, detain, divert, treat, release).

2) Pretrial status decisions (release on recognizance, release on unsecured or secured bond, release with supervision conditions, detain, respond to noncompliance, reassess supervision conditions).

3) Diversion and deferred prosecution decisions.

4) Charging decisions (charge, dismiss).

5) Plea decisions (plea terms).

6) Sentencing decisions (sentence type, length, terms and conditions).

7) Local and state institutional intervention decisions (security level, housing placement, behavior change interventions).

An example of a tool used in the arrest decision in point 1 above is the Police Diversion Screening Tool developed by the Criminal Justice Lab at NYU Law (Criminal Justice Lab, NYU Law, n.d.). The tool, developed in conjunction with the Indianapolis Metro Police Department, allows for police in the field to divert individuals with mental health or substance abuse issues away from the criminal justice system and directly into an appropriate mental health facility. In addition to reducing congestion in local jails, the tool allows for appropriate treatment for those who need it, thereby possibly reducing recidivism.

Some of these decisions must be made within a noticeably short time frame, such as an arrest decision. Even in this case though, the almost instantaneous availability of information about the potential arrestee and the availability of fast evaluation algorithms make the decision easier.

In the case of decisions associated with federal laws, some of these laws are passed to repeal or amend old laws, and some are completely new. The number of federal laws in existence is almost impossible to know for sure, but an exhaustive study done in the 1980s counted approximately 3,000 different criminal offenses (Cali, 2013).

In addition to the types of decisions peculiar to criminal justice systems, there are the decisions associated with most large-scale systems, including those related to funding levels, locations, allocations, capacities, staffing, scheduling, routing, privatization, etc. Specifically, we are referring to questions like:

- Where should a new prison be located and what should be its capacity?
- Should a prison be privatized (e.g., operated by a private company rather than the state government)?
- Should the budget for a police department be increased by 20%, reduced by 20%, or eliminated altogether?
- Should an increased level of policing be done for a particular area of the city?

In a more general sense, the potential decisions associated with criminal justice systems might be categorized as follows: (1) modification, elimination, or institution of new state or federal laws/statutes or regulations; (2) modification, elimination, or institution of new policies/procedures; and (3) the addition or deletion of resources (law enforcement, judges, prosecutors, probation officers, prison/jail staff, etc.) or facilities.

2.2.4 Analytics and Operations Research

Many of the decisions associated with criminal justice systems can be addressed through the use of **analytics**, and a closely related field,

operations research. Analytics can be defined as *the scientific process of transforming data into insights for the purpose of making better decisions* (Best Definition of Analytics, n.d.). The emphasis in analytics is on making decisions, which happens to be the emphasis in this book.

Analytics is often categorized according to three types: descriptive analytics, predictive analytics, and prescriptive analytics (Operations Research and Analytics, n.d.). Prescriptive analytics, the focus of this book, yields guidance for making decisions. This guidance is accomplished using **models** or representations of systems.

Many decision makers employ a mental model of the situation when deciding. Typically, in these mental models, the decision maker will mentally assess the various respective outcomes associated with the alternative decisions and then select the decision which gives the best predicted outcome. These outcomes can correspond to the values associated with the various performance measures for the situation, so the decision maker will need to trade-off among these conflicting performance measures.

For example, a state could have a decision to make as to which private company will be selected to operate a new prison. (This issue of privatization of prisons is discussed in Chapter 3.) In making this decision, the state would naturally consider several conflicting performance measures, including those related to cost, safety, security, living conditions, rehabilitation programs, etc. Naturally, trade-offs would need to be made between cost and several of the quality-oriented measures.

In analytics and operations research, we use models which are developed and experimented with on a computer. These models can be any of several different types, including a decision tree, an optimization model, a simulation model, etc.

One of the advantages of such computerized models is that they allow for the experimentation with different alternatives without disturbing the original system. For example, consider the First Step Act, signed into law in December of 2018, and discussed in some detail in Chapter 4 of this book. This law had several parameters, including the funding of **$244 million** for recidivism-reducing programs for pre-release custody inmates and the earning of **10 days** of early-release credit for every 30 days of successful engagement in effective activities by the inmate. With an appropriate simulation model of the federal prison system, one could easily experiment with these and other values for the two parameters. Such a model would allow for the representation of many years of simulated time with a few seconds of actual time, thereby allowing for the simulation of many policies.

In this book we focus on the use of some methodologies from the area of multiple objective analytics; these methodologies include multiattribute value functions and multiattribute utility functions. The reason for the focus on multiple objectives is that the decision situations in criminal justice systems naturally lend themselves to these methods, as will be discussed in the next subsection.

2.2.5 Complexities Which Make Decisions Difficult for Criminal Justice Systems

There are several complexities associated with criminal justice systems which result in major difficulties for their design and operation. These complexities include the following:

- Criminal justice systems have many interacting subsystems, including those related to policing, courts, and corrections.
- Criminal justice systems have complex interactions with other systems, such as those related to social welfare, mental and physical health, education, etc.
- Criminal justice systems have many decision makers and stakeholders for their various decision situations, resulting in the existence of several conflicting performance measures.
- Criminal justice systems behave in a dynamic fashion, resulting from changing demographics, economics, laws, and morals.
- Criminal justice systems behave in a probabilistic manner due to many reasons, not the least of which has to do with the uncertainty of human behavior.
- Making changes to criminal justice systems, through for example new laws, is typically accomplished through an often highly partisan political environment.
- Good data for making decisions about criminal justice systems is often difficult to obtain.

As mentioned earlier, the three major subsystems of a criminal justice system are law enforcement, the court subsystem, and the prison/jail subsystem. Each of these subsystems can be further subdivided. For example, the prison/jail system can be divided into county jail systems, state prisons, federal prisons, probation services, etc. A change which initially affects one system/subsystem can subsequently affect others, which makes the accurate prediction of the effect on the overall system difficult to accomplish; hence, the need for multiple objective analytics.

As an example of a law that was enacted to change the operation of one part of a criminal justice system, but then resulted in (perhaps unanticipated) changes to other parts, consider the habitual offender law (often called the "three-strikes law"). This law has been enacted in several states over the last 30 years. There are many versions of the law; one version states that a severe violent felony, coupled with two previous convictions will result in a mandatory life sentence for the offender. Auerhahn (2008) illustrated how the three-strikes law in California resulted in a chain reaction in which fewer of the accused with two previous convictions entered guilty pleas, resulting in a much more congested jail and court system in California. This in turn resulted in court orders to cap the jail population, resulting in early release of many prisoners from county jails.

Criminal justice systems interface with and are greatly affected by other systems: social welfare systems, educational systems, economic systems, mental health systems, and health care systems in general. Addressing problems in these other systems can result in benefits to criminal justice systems. Not addressing many of the problems in these other systems can result in an overcrowded criminal justice system. For example, a poor educational system can lead to an increased dropout rate, which in turn can lead to increased criminal behavior.

Because there are typically many decision makers and stakeholders for most important decision situations of a criminal justice system, there can be different objectives and corresponding performance measures. The example discussed earlier in this chapter, involving which private company should be chosen to operate a state prison, had objectives related to cost, safety, security, living conditions, and rehabilitation programs. Making a decision in this situation obviously involves making trade-offs between cost and any of the objectives related to quality.

Changes to a criminal justice system affect the system over time; in addition, outside influences such as demographic characteristics of the population change over time, hence the dynamic nature of these systems.

The fact that criminal justice systems are dynamic in nature arises from the fact that demographics, laws, social customs, economic conditions, technology, and other important aspects of the human condition are changing over time. Certainly, decisions that have a strategic or long-term effect, such as the enactment of federal laws, must, at least implicitly, forecast these changing conditions in order to gage the outcome associated with the enactment.

Since system changes result in effects that occur in the future and which also depend on human behavior, the effects will typically have a large amount of inherent uncertainty. For example, with respect to early release programs, there will be uncertainty as to whether a released person will commit more crimes. Hence, modeling approaches which accurately represent uncertainties in the behavior of persons released from prison are important to employ.

Human behavior is difficult to exactly predict, but one may be able to predict it in a probabilistic fashion. For example, a defendant released on bail or on his/her own recognizance may or may not appear for his/her court date and may or may not commit crime(s) prior to trial. However, by considering various characteristics of the situation and the defendant, a probability distribution over the various actions of the defendant may be assessed. Such a probability distribution may be input to a model which would aid the judge in deciding about the pre-trial disposition of the defendant.

The political nature of decisions associated with criminal justice systems is exacerbated by the interpretation of the data analysis associated with these systems. For example, Barnett (1988) provides three examples of misinterpretation of data involving (1) racism and the death penalty in Georgia; (2) the hypothetical increase in crime that would occur if all prisoners were released from jail or prison; and (3) whether defendants who were arrested for violence, would be more likely than other defendants (i.e., those arrested, but not for violent behavior) to engage in pretrial misconduct if released on their own recognizance.

The first example discussed by Barnett involved a Supreme Court case in the state of Georgia, discussed in an article in the *New York Times*. The contention was, as stated in the *Times* article and in Barnett's paper, that:

> *other things being equal as statisticians can make them, someone who killed a white person in Georgia was four times as likely to receive a death sentence as someone who had killed a black.*

Barnett showed how this statement could be easily misinterpreted, given the facts of the study. Even though the examples cited by Barnett were from more than 30 years ago, the practice of misinterpreting data for political gain continues to date. Examples can be seen in Chapter 4 of this book involving the recent discussion of the Violent Crime Control and Law Enforcement Act of 1994.

Finally, with respect to the lack of good data, as discussed in Chapter 1, many of the data collection systems involve self-reporting which is often inaccurate in nature.

The seven complexities discussed above make the prediction of the effect of a change to a criminal justice system, such as a new law, difficult to accomplish. In turn, this difficulty makes the choice of a best alternative (i.e., optimization) a difficult process. This difficulty occurs no matter what types of analyses are employed for the decision process.

Many of the complexities discussed above are typically found in what are termed ill-structured problems (Simon, 1960). Hence, it is important to consider problem structuring methods when addressing decision situations for criminal justice systems.

In the next section we will discuss four categories of methods that can aid in addressing several of the difficulties presented above. These are (1) problem structuring methods, (2) methods for generating the multiple performance measures for a decision situation, (3) methods for representing the preferences of a decision maker over multiple performance measures, and (4) simulation modeling and analysis as a way to represent the dynamics and probabilistic nature of a system.

2.3 ANALYTIC METHODOLOGIES FOR DEALING WITH CRIMINAL JUSTICE SYSTEMS

2.3.1 Problem Structuring Methods

A problem can be defined as a gap between a current and a desired state of affairs. In this context, state of affairs refer to a set of values assigned to some respective performance measures. For example, with respect to the criminal justice system one problem could be defined as:

> 67% of prisoners released from incarceration are re-arrested within three years (Durose, Cooper, and Snyder (2014)). We would like that figure to be at most 20%.

The gap in this case is between the 67% and 20% at a maximum. This problem is one that might be considered as an ill-structured problem as opposed to a well-structured or semi-structured problem.

As discussed by Simon (1960) problems fall along a range of complexity, from ill-structured to well-structured. Ill-structured problems, such as the one above involving prisoner recidivism, are those for which we have many difficult-to-identify stakeholders and decision makers, and, as a result, there are many conflicting objectives. In addition, good data for modeling these problems are often difficult to obtain, and good alternative solutions are not immediately obvious. Finally, the root causes are often difficult to identify.

Typically, ill-structured problems are embedded within a whole network of problems, termed *messes* by Ackoff (1979). Many of the methodologies associated with problem structuring involve the identification of these networks. Identifying the network of problems allows one to more easily establish the root causes of the problem, the relevant stakeholders and decision makers, good alternative solutions, sources of uncertainty and risk, and important systems that are connected to the system in which the initial problem is embedded.

Some of the important problem structuring, and related, methodologies include cognitive mapping (Eden and Ackermann, 2004), breakthrough thinking (Nadler and Hibino, 1990), the Kepner and Tregoe Method (Kepner and Tregoe, 1981), the Delphi Method (Linstone and Turoff, 1975), and the Why-What's Stopping (WWS) problem structuring heuristic (Basadur et al. (1994) and Ellspermann et al. 2007)).

These problem structuring methods typically involve the efforts of a group of participants, consisting of an analyst or facilitator, stakeholders, and decision makers with different perspectives on the problem. This group will work in an interactive fashion, to generate the problem structuring output. The approaches also typically involve "divergent thinking" (an example of which is brainstorming) to generate the output from the process.

The WWS heuristic starts from a single problem statement to generate a whole network of problem statements, with various perspectives on the problem. This network contains problem statements which basically correspond to alternative solutions to the problem. As noted above, group interaction involves a facilitator and a variety of stakeholders with differing perspectives. The discussion in the group is supposed to be non-critical in nature; that is, no problem statement suggested by a group member is to be criticized by another group member.

The output from the WWS heuristic will be a two-dimensional problem map, where each node in the map corresponds to a problem statement. An arc pointing upward from one problem statement to another corresponds to a problem statement which is more specific in nature leading to a problem statement which is more general in nature. Correspondingly, an arc pointing downwards from one problem statement to another corresponds to a more general problem statement leading to a more specific problem statement.

2.3.1.1 An Illustrative Example of Problem Structuring for Juvenile Crime

The problem statements generated by the WWS heuristic start with the phrase "How might we/I..." to provide the process with an optimistic perspective. For example, a problem statement could be given as:

How might we greatly reduce juvenile crime in our city?

Addressing a problem such as this is certainly important given the facts that (1) the great majority of adult criminals start their criminal behavior as juveniles and (2) in 2018, juvenile jurisdiction courts disposed of approximately 744,500 cases, of which 232,400 involved offenses to a person (homicide, rape, robbery, simple assault, aggravated assault, etc.), 225,900 involved property offenses, 101,000 involved drug law violations, and 185,100 involved public order offenses. Though 744,500 is a large number of cases, it does represent a decrease nationally over time. In particular, the total number of annual cases had decreased by 5% from 2017 to 2018, 48% from 2009 to 2018, and 35% from 1985 to 2018 (Statistical Briefing Book, n.d.). Even though there has been a decrease on the national level, local communities have seen increases.

Following this initial problem statement, the facilitator can ask either of two questions:

1) Why do we want to greatly reduce juvenile crime in our city? or
2) What's stopping us from greatly reducing juvenile crime in our city?

The answers to the first question (why) will generate several problem statements of a more general nature than the original problem statement. These more general problem statements can allow us to view the original problem from a different perspective.

The answers to the second question (what's stopping) will allow us to generate several problem statements of a more specific nature.

These more specific problem statements may very well correspond to specific actions that can be taken as solutions to the problem.

Answers to the question: "Why do we want to greatly reduce juvenile crime in our city?" might be:

- We want our young people to lead productive lives.
- We want our young people to serve as good role models for others.
- We want to reduce the costs associated with crime in our city.
- We want to have safe shopping areas, safe residential areas, and safe parks in our city.
- We want to create a good business climate in our city.

Answers to the question: "What's stopping us from greatly reducing juvenile crime in our city?" might be:

- Many of our young people do not have a sense of belonging to something meaningful.
- Many of our young people drop out of high school.
- Many of our young people have poor role models.
- Many of our young people are living in poverty.
- Many of our young people are substance abusers.
- Many of our young people live in dysfunctional single-parent homes.
- Many of our young people live in homes where one or both parents are substance abusers.
- Our police are not being used efficiently.
- We need to increase funding for law enforcement.
- We need to improve rehabilitation programs for young people who are convicted of crimes.
- We need to reduce or eliminate harmful effects of the internet.
- Many of our young people have personality dysfunctions or emotional problems.
- We need to have a way to keep dangerous juvenile offenders off the streets.
- We need more alternative schools.

Of course, there are many other statements that we could provide as answers to these questions. Once we have these statements, we re-phrase them into "How might we..." statements, such as: "How might we have our young people lead productive lives?" and "How might we provide young people with better role models?". Each of these "How might we…" problem statements are placed in respective rectangles or nodes as part of a diagram. An arc is placed in the diagram from one node to another with an upward orientation if the latter node represents a problem statement which is an answer to the "why question" for the former node's problem statement; the arc has a downward orientation if the latter node represents a problem statement which is an answer to the "what's stopping" question for the former node's problem statement.

At this point we would have a problem network with three levels of problems. The top level would have five problem statements, the middle level would have one problem statement, and the bottom level would have fourteen problem statements. Note that the higher-level problem statements address the issue from a broader, more general perspective; the lower-level problem statements address the issue from a more specific perspective.

Continuing with the problem network, we could expand upward or downward from any node (problem statement) in the network. As we expand downward, we eventually get to problem statements which represent alternative solutions to the original problem. Expanding downward from one of the higher-level problem statements could very well lead to solutions which would not have been obvious from expanding downward from the original problem statement.

See Figure 2.1 for a problem network that could have been developed with a little more effort from the initial work discussed above. Note that expanding down from the problem statement "How might we reduce the costs associated with crime in our city?" yields the problem statement: "How might we get our businesses to have better security measures?". This latter problem statement could lead to **solutions** for better security measures for businesses, which may not have been an obvious solution derived from the initial problem statement.

It's probably true that someone familiar with the area of juvenile crime would be able to develop as much or more content as is contained in Figure 2.1. However, if the WWS heuristic were applied in an actual setting involving teachers, business owners, students, parents, school administrators, police, youth counselors, etc., then the opportunity to discuss priorities among the various participants would be available. The advantages of the various lower-level problem statements (solutions) could be discussed in more detail.

For additional discussion of the WWS heuristic, see pages 29–36 of Evans (2017) or Basadur et al. (1994).

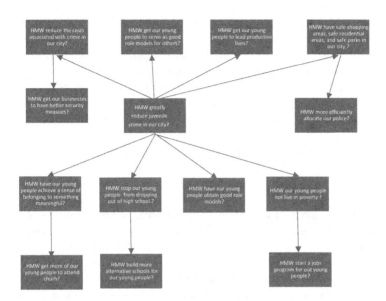

Figure 2.1 A partial WWS problem network for the problem: "how might we greatly reduce juvenile crime in our city".

2.3.2 Generating Multiple Objectives and Multiple Performance Measures for a Decision Situation

Making decisions about the design and operation of criminal justice systems requires the consideration of multiple, conflicting performance measures. For example, in a decision by a state as to whether to privatize its prisons, performance measures related to cost, security, effectiveness of programs for rehabilitation, quality of health care, among others, needs to be considered.

In the vernacular of decision analysis, these performance measures are called **attributes**. An attribute is associated with an objective and can be thought of as a measure of that respective objective. For example, an objective that one might consider in a decision involving privatization of prisons, mentioned in the previous paragraph, would be to maximize the effectiveness of rehabilitation programs. An attribute that one might consider for this objective would be the percentage of prisoners who are rearrested within three years of release from prison. (Note that we want to maximize the objective and minimize the attribute value in this case.) An objective can always be stated as a phrase that starts with one of the following words: maximize, minimize, or optimize.

In many situations, the set of objectives and associated attributes for a decision situation can be presented in the form of a hierarchy, with the more general objectives (e.g., optimize well-being of prisoners) toward the top of the hierarchy and the more specific objectives toward the bottom of the hierarchy (e.g., minimize recidivism).

2.3.2.1 An Illustrative Example: Determining Objectives and Performance Measures for Setting Up a Syringe Exchange Program

Consider a decision situation involving the institution of a syringe exchange program (SEP) for an urban area. (These programs are also called needle exchange programs (NEPs), or syringe service programs (SSPs).) Through the use of storefronts, mobile vans, pharmacies, outreach programs, and syringe vending machines, these programs allow intravenous drug users to (1) exchange used syringes for clean syringes; (2) obtain condoms, cookers, purified water, and bleach; and (3) obtain counseling services and educational material for their drug addiction problems.

The basic idea is that intravenous drug users, especially those who are homeless or plagued with poverty, are not overly concerned with the obtainment of clean syringes when they inject themselves. As a result, they are at much greater risk for contracting the human immunodeficiency virus (HIV), the hepatitis B virus (HBV), or the hepatitis C virus (HCV). HIV causes AIDS. For example, in 2013 in the United States, 3,096 of the 47,352 diagnoses of HIV infection were attributed to injection drug use. In addition to the obvious problems associated with the health of the individuals involved, there are the economic problems associated with treating the diseases. For example, it has been estimated that the lifetime treatment for HIV is $379,668 in 2010 dollars (Access to clean syringes, n.d.). Much of this cost is paid with public funds. See Syringe Service Programs (SSPs) FAQ (2019) for additional information about these programs.

A review of 15 inquiries involving the analysis of SEPs found that these programs were associated with large decreases in HIV and HCV infections (Abdul-Quader et al, 2013). In particular, it was found that the establishment of New York City's SEP resulted in an increase in annual syringe exchanges from 250,000 to 3,000,000 during the period from 1990 to 2002; this corresponded to a decrease in the prevalence of HIV in injection drug users from 50% to 17% during this period (Des Jarlais et al, 2005).

Note that there are other means for increasing the use of clean syringes, including the allowance of nonprescription sale of syringes and needles, and the introduction of statutes which eliminate syringes from being classified as drug paraphernalia.

The effects associated with SEPs are not all good of course. (Almost any changes made to a criminal justice system through new laws, programs, procedures, etc., will have both good and bad points, which is one of the essential ideas of this book.) For example, the establishment of SEPs has a high dollar cost for staffing, facilities, and supplies. Second, many people look at the establishment of an SEP as a program that can encourage drug use. Third, many government officials, especially those who are known for their conservatism, consider the political backlash associated with the implementation of SEPs as being bad for them. For example, Mike Pence, then Governor of Indiana, initially resisted the establishment of SEPs in the state over the objections of health officials, before finally relenting (Demko, 2016). These government officials will often base their objections on religious or moral grounds and will cite the possible encouragement of drug use because of the programs (Hedger, 2017).

Finally, many neighborhoods object to the establishment of SEPs in their respective localities. For example, there are usually objections to the relevant facilities being located "too close" to a school (Hinton, 2019) and people who live in the neighborhood often feel unsafe due to the inflow of addicts (Ackerman, 2019).

The decision situation with which we are concerned here involves things like whether an SEP should be implemented in an area, as well as the design and operation of the program. Design and operation imply things like where facilities are located, the capacities of the facilities, routings for mobile facilities, and rules by which the program will operate. As an example of an operating rule, a neighborhood group in Asheville, North Carolina, wanted all participants in an SEP to be required to participate in a rehabilitation program. The program operators resisted this suggestion since their experience indicated that an environment of non-coercion would be necessary to attract participants (Hinton, 2019).

There are several (often complementary) techniques available for generating a hierarchy of objectives and corresponding attributes for a decision situation. For example, Buede (1986) suggests a top–down approach for strategic decisions and a bottom–up approach for tactical or operation decisions.

As the names indicate, a top–down approach would start with one general or strategic objective at the top of the hierarchy, like:

> Design the syringe exchange program so that the benefits of all stakeholders are optimized.

Note that we are referring to benefits in this top-level objective in a very general way such that, for example, a cost would be a "negative benefit". The stakeholders in this case would be: (1) the potential clients of the SEP, (2) neighborhood residents, (3) taxpayers, (4) government officials, and (5) staff of the SEP. Note that one could further divide some of these stakeholder groups into sub-groups to define additional stakeholders; for example, neighborhood residents could be subdivided into schools, residents living in houses/apartments, business owners, etc.

A bottom–up approach would start with the different alternatives for an SEP and expand upward based upon the differences in the outcomes associated with the alternatives. For example, the different alternatives could be (1) no SEP, (2) SEP with Design 1, (3) SEP with Design 2, and so on. As noted above, a design could imply location(s) of facilities, routes for mobile facilities, specific staffing levels (which imply specific capacities for the facilities), rules under which the program would operate, etc. One of the differences in these alternatives would be the number of clients served on an annual basis. Hence, a lower-level objective for the hierarchy could be "maximize the number of clients served on an annual basis", with a corresponding attribute of "number of clients served on an annual basis".

One can also use any **device** from a list of devices for generating objectives for a decision situation (page 57, Keeney, 1992). Some of these devices suggested by Keeney include: a wish list, problems and shortcomings, different perspectives, consequences, goals, constraints, and guidelines, among other things. As an example, one of the guidelines that might be specified by an addiction specialist would be to not have too much urging of the clients by the staff with respect to entering a rehabilitation program. Hence, an objective would be to minimize coercion of clients with respect to entering a rehabilitation program.

Another approach that can be useful in generating a hierarchy of objectives is the use of **specification** and **means-ends**. For example, in expanding downward from a higher-level objective, one could **specify** what is meant by the higher-level objective, or one could ask what is the **means** by which that higher-level objective is achieved. For example, one might ask "What is the means by which we can minimize troubles caused to the schools?", with the answer being "maximize the distance from any school to its closest SEP facility".

On the other hand, in moving up the hierarchy (from a lower-level objective to a higher-level objective) one might ask "To what **end** do we want to achieve this lower level objective?".

Bond, Carlson, and Keeney (2010) suggest a two-stage approach for generating the hierarchy. The first stage involves having a group of diverse stakeholders for the decision situation brainstorm to generate an unstructured group of objectives. This is followed by a stage in which the "analyst" provides structure by organizing the objectives into similar groups. Then the first stage is repeated in some sense by having the stakeholders add additional objectives by thinking more intensely about the individual groups of objectives. The analyst may set a goal such as adding 50% more objectives for the second stage of the process. It may be useful for the stakeholders to employ some of the methods espoused by value-focused brainstorming (Keeney, 2012).

Additional approaches for generating objectives for decision situations include examination of the literature, categorization of stakeholders, fundamental vs. means objectives (Clemen and Reilly, 2013), and questions for moving up/down the hierarchy (Clemen and Reilly, 2013). For additional discussion on these and the other approaches discussed in the preceding paragraphs, see Evans (2017).

Using one or more of the approaches discussed above, we might formulate an initial part of the hierarchy for the decision situation involving the establishment of an SEP as shown in Figure 2.2.

Expanding further on the initial hierarchy of Figure 2.2, under the "Minimize cost" node we could have two objectives:

O1: Minimize initial cost and
O2: Minimize monthly recurring costs.

Under "Minimize troubles caused to the neighborhood" we could have the objectives of:

O3: Minimize troubles caused to schools,
O4: Minimize troubles caused to residents, and
O5: Minimize troubles caused to businesses.

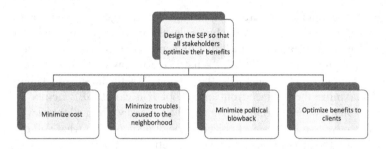

Figure 2.2 A partial hierarchy of objectives for a decision situation involving the establishment of a Syringe Exchange Program.

Under "Optimize benefits to clients", we could have the objectives of:

O6: Maximize the number of clients served on an annual basis,

O7: Minimize average travel distance for clients,

O8: Maximize the number of HIV infections averted on an annual basis,

O9: Maximize the number of HBV infections averted on an annual basis,

O10: Maximize the number of HCV infections averted on an annual basis, and

O11: Maximize the number of addicts rehabilitated on an annual basis.

We could consider the above objectives: O1 through O11, as lowest-level objectives, or we could consider expanding on at least some of these objectives further. For example, for O3, minimize troubles caused to schools, we could ask the question: "What is the means by which we can minimize troubles caused to schools?". The answer might be: "maximize the sum of the distances from each school to its closest SEP facility"; alternatively, the answer might be "maximize the average of these distances".

Finally, for this example, we would use one of the objectives already shown in Figure 2.2 as our last "lowest- level objective":

O12: Minimize political blowback.

One could continue to expand on objective O12. For example, an objective that would be a lower-level objective, related to O12, would be "minimize probability of not being re-elected" (certainly crass in nature, but corresponding to real life). However, for this example, we will stay with the objectives stated above, O1, O2, O3, ..., O12.

Now, for each of our lowest-level objectives we need to define an attribute to measure that objective. Attributes can either be **natural** (sometimes called **quantitative**) or **constructed** (sometimes called **qualitative**) in nature. Usually, whether one uses a natural attribute or a constructed attribute will be obvious. For the objectives defined above, O1, O2, O6, O7, O8, O9, O10, and O11 will require obvious natural attributes, as defined below:

O1 attribute: initial cost for SEP in dollars,
O2 attribute: monthly recurring costs for SEP in dollars,
O6 attribute: number of clients served on an annual basis,
O7 attribute: average travel distance for clients,
O8 attribute: number of HIV infections averted on an annual basis,
O9 attribute: number of HBV infections averted on an annual basis,
O10 attribute: number of HCV infections averted on an annual basis, and
O11 attribute: number of adults rehabilitated on an annual basis.

Note that each of these attributes can be assigned straightforward numerical values.

The attributes for objectives O1 and O2 would be relatively easy to compute values for any particular SEP design. For the attributes of the other objectives (O6 through O11), predicting values for any particular design (implied by the number and capacities of facilities, routes for mobile facilities, operating procedures, etc.) would be more difficult. At least some of these predictions might be probabilistic in nature, but the predictions could be accomplished through the use of sophisticated mathematical models, or through the use of expert opinion. Examples of the types of mathematical models that can be used to make these predictions are found in Kaplan and O'Keefe (1993).

The attributes for objectives O3, O4, O5, and O12 would be constructed attributes, or qualitative in nature. Associated with each of these attributes would be a qualitative scale, say based on the values of 0, 1, 2, 3, 4, or 5, with 0 being the worst possible value, and 5 being the best possible value:

O3 attribute: troubles caused to schools on a scale of 0–5, with 0 being the worst possible value and 5 being the best possible value,

O4 attribute: troubles caused to residents on a scale of 0–5, with 0 being the worst possible value and 5 being the best possible value,

O5 attribute: troubles caused to business on a scale of 0–5, with 0 being the worst possible value and 5 being the best possible value, and

O12 attribute: amount of political blowback on a scale of 0–5, with 0 being the worst possible value and 5 being the best possible value.

Note that the scales used do not necessarily have to range from 0 to 5, with the worst possible value in each case being 0, and the best possible value being 5. The ranges could be 0–4, 0–10, 1–5, or something else. However, something on the order of 0–5 typically works well. In addition, we could let the best possible value be 0 and the worst possible value be 5 if we wanted, if the succeeding activities with respect to the modeling of preferences represented this.

Associated with each of the numbers on the subjective scales should be a description of what is meant by those numbers, on an operational level. For example, a "5" on the O4 attribute scale might correspond to something like:

> a very small minority, if any, of the students in the area will even realize that there is an SEP in the area.

The key thing is to make as clear as possible the meaning of the number on the subjective scale for the attribute.

A different approach than the ones described thus far, for determining a good set of objectives and attributes for a decision situation, is described by Keeney and Gregory (2005). Their process suggests that, as one is expanding downward in a hierarchy of objectives, if one natural attribute is an obvious choice for an objective, then it should be used; the expansion from that objective will then stop. If that is not the case, then one might consider several alternative natural attributes and choose the best of all of them. If neither of these actions is reasonable, then one might consider expanding the relevant objective into a set of lower-level (or component) objectives, and then repeat the process for each of these objectives.

Judging the quality of a set of objectives and corresponding attributes for a decision situation is something of a subjective process. This judging of the quality of a set of objectives, or of attributes usually involves determining whether the set has a particular group of characteristics. For example, Keeney (1992, pages 82–86) suggests that a group of fundamental objectives for a decision situation has nine characteristics, namely that they be essential, controllable, complete, measurable, operational, decomposable, nonredundant, concise, and understandable.

In a similar fashion, Keeney and Gregory (2005) suggest that a set of attributes for a decision situation should have the following characteristics, as a whole: unambiguous, comprehensive, direct, operational, and understandable. The meanings of these characteristics should be clear, but as an example, consider the characteristic of being operational. This means that we can make reasonable predictions of the values of the attributes for any alternative under consideration. In particular, in the case of the SEP decision situation, the model used to predict the annual number of HIV infections averted must be accurate, and the decision makers involved with the choice of an alternative must be able to make trade-offs between this performance measure and the others; in particular, the decision makers must be able to give an answer to a question like: "How much would you be willing to pay in terms of annual operating cost in order to avert ten annual HIV infections?".

2.3.3 Cost–Benefit Analysis

Cost–benefit analysis (CBA) is a methodology which evaluates each alternative in a decision situation by categorizing all of the effects of the alternative as either a cost or as a benefit, attaching a dollar value to each, and then seeing if the dollar value of the benefits is greater than the dollar value of the costs. Future costs and benefits are discounted with an appropriate interest rate.

Appropriate modeling techniques must often be used to accurately predict at least some of the costs and benefits, especially if they correspond to effects scheduled to happen far into the future. In addition, some of the costs and benefits might correspond to intangibles, which can be difficult to calculate. Examples of these intangibles with respect to criminal justice systems would be things such as the costs associated with the fear of crime, the death/injury of an individual through an assault, the detrimental effects on the family of someone sentenced to prison.

An approach for applying CBA in criminal justice in general is embodied in the Manning Cost–Benefit Tool (MCBT) (Manning et al., 2016). This approach basically involves having the user input data to spreadsheets; the data consists of a variety of inputs corresponding to some type of intervention in the criminal justice system. One of the major contributions of the MCBT is the enumeration of most of the categories of the various costs and benefits of an intervention of this type. For example, there are the major categories of (1) costs in anticipation of crime, (2) costs as a consequence of crime, and (3) costs in response to crime. These are divided into the sub-categories of homicide, serious wounding, sexual offenses, common assault, robbery, etc. The tool has as one of its inputs a discount rate, which allows for the computation of the net present value of costs and of benefits. In a follow-on article, Manning et al (2018) suggest the use of a conceptual tool called Smart MCBT that uses machine-learning techniques to aid the user in the input of the vast amount of data needed for the use of the tool.

The state of Washington has performed cost–benefit analyses of several of its programs and policies. (See the Washington State Institute for Public Policy (n.d.) and Aos and Drake (2010)). Their probabilistic approach to the analyses allowed for an estimate of the probability that the benefit would exceed the cost, as well as an estimate of the net present value of the benefits minus the costs, and a benefit to cost ratio. For example, their analysis of an employment counseling and job training program for transitioning from incarceration into the community yielded a net present value of $43,502, a benefit to cost ratio of 18.21, and a probability associated with the benefits exceeding the costs of .88.

Fass and Pi (2002) used simulation and CBA to investigate juvenile crime in Dallas County, Texas. Their investigation ascertained that more severe sentencing for juvenile crimes could avert some offenses, but that the cost for implementing the actions was much greater than the benefits.

Finally, Zarkin et al (2015) noted that 50% of those incarcerated in state prisons meet the standards for diagnosis of drug dependence. They developed a simulation model to look at the net benefits of diversion from reincarceration to community-based substance abuse treatment and discovered that the benefits far outweighed the costs.

2.3.4 Multiattribute Value Functions and Multiattribute Utility Functions

Once a set of attributes for a decision situation has been properly identified, one can develop a method for ranking outcomes which correspond to a set of values assigned to these attributes. These outcomes would be associated with the respective alternatives under consideration. For example, one may be considering various policies corresponding to the hiring of additional police; each policy can correspond to an outcome corresponding to additional costs and higher levels of arrests.

In some cases, these outcomes may be deterministic in nature, and in other cases they may be probabilistic. One way to rank these outcomes if they are deterministic (probabilistic) in nature is through the use of a multiattribute value function (multiattribute utility function).

A **multiattribute value function** is one which maps from the outcome space to the space of real numbers. It is used to represent the preference structure of a decision maker over the outcome space of multiple attributes. Suppose that we use the following notation, corresponding to that used in Evans (2017):

n is the number of alternatives under consideration,
A_i represents alternative i for i = 1, …, n,
p is the number of attributes that we have under consideration,
X_k represents attribute k for k = 1, …, p, and
$x_k(A_i)$ is the attribute value associated with attribute X_k for alternative A_i.

In most cases involving the use of a multiattribute value function, the function v will be what is termed a scaled, additive function, as shown below:

$$v\left(x_1, x_2, x_3, …, x_p\right) = \sum_{k=1}^{p} c_k v_k \left(x_k\right),$$

where:

$v_k\left(x_k^b\right) = 1$, and $v_k\left(x_k^w\right) = 0$ for k = 1, …, p where x_k^b is the best possible value for the kth attribute and x_k^w is the worst possible value for the kth attribute,
$0 < c_k < 1$, for k = 1, …, p, and

$$\sum_{k=1}^{p} c_k = 1.$$

Note that the v_k for k = 1, …p, are called individual attribute value functions. These are not necessarily linear. In addition, the slope of an individual attribute value function will be positive if the attribute is one that should be maximized, and negative if the attribute is one that should be minimized.

Determining a multiattribute value function for a decision maker is called the assessment process. There are four basic sequential steps for this assessment process: (1) determine the best and worst values for each attribute, (2) determine whether the decision maker's preferences correspond to a multiattribute value function with an additive form, as shown above, (3) determine the individual attribute value functions, $v_1, \ldots, v_p$, through answers obtained from the decision maker about his/her preferences over the outcome space, and (4) determine the scaling constants: $c_1, c_2, \ldots, c_p$, again through the answers obtained from questions posed to the decision maker.

Within each of the four steps, there are one or more approaches that can be used. The exact assessment process is beyond the scope of this book, but one can view any of several sources to obtain additional information about assessment (pages 94–112 of Evans (2017), or pages 82–125 of Keeney and Raiffa (1993)).

The reader can view an example of the use of a multiattribute value function in Chapter 3. This hypothetical example scores and ranks proposals from private companies for the operation of a prison.

A **multiattribute utility function** is also typically scaled to provide a value between 0 and 1 when acceptable values for attributes are input as independent variables for the function. The main difference between this function and a multiattribute value function however is that a multiattribute utility function will allow for the ranking of **probabilistic** outcomes, through the criterion of maximization of expected utility.

As with a multiattribute value function, a multiattribute utility function represents the preferences of a specific decision maker. In addition, typically, when applied in specific situations, these functions will assume any of several different specific forms, e.g., additive, multiplicative, or multilinear (see pages 236–263 of Evans (2017)).

The process of assessment of a multiattribute utility function, as with a multiattribute value function, also involves the decision maker(s) answering questions about their preferences over probabilistic outcomes defined over multiple attributes.

2.3.4.1 An Illustrative Example: Using a Multiattribute Utility Function for the Decision of Whether to Use a Court-Appointed Attorney

Let us consider a hypothetical example where one could use a multiattribute utility function. This example illustrates how one can employ a multiattribute utility function.

Suppose a defendant has a decision to make. The defendant can use a court-appointed attorney to handle the felony case (decision 1), or the defendant can hire an attorney (decision 2), at a cost of $3500, to handle the case (How much are attorney fees?, n.d.). If the defendant uses the court-appointed attorney, then the probability of being convicted and serving two years in a state prison is .7, and the probability of being found innocent and therefore set free is .3. On the other hand, if the defendant hires the (expensive) attorney at a cost of $3500, the probability of being found guilty is .4, and if found guilty, there is a .5 probability of serving two years in prison, and a .5 probability of serving one year in jail.

From the description of the problem there are basically two attributes with which the defendant will be concerned:

X_1 = cost in thousands of dollars and
X_2 = number of months spent in jail/prison.

The outcome associated with each decision will be probabilistic in nature. For example, if the defendant decides to use the court-appointed attorney, the outcome will be:

X_1: 0 with certainty,
X_2: 0 with .3 probability and 24 with .7 probability.

The outcome associated with the second alternative decision of hiring the attorney is:

X_1: 3.5 with certainty,
X_2: 0 with .6 probability, 12 with .2 probability, and 24 with .2 probability.

Now, the best possible values for X_1 and X_2, denoted as x_1^b and x_2^b, respectively, are 0 and 0. The worst possible values for X_1 and X_2, denoted as x_1^w and x_2^w, respectively, are 3.5 and 24.

Let's suppose that the defendant (who is the decision maker in this case) has a multiplicative utility function over the two attributes of concern. This multiplicative function, denoted as $u(x_1, x_2)$ could be written as follows:

$$u(x_1, x_2) = w_1 u_1(x_1) + w_2 u_2(x_2) + w w_1 w_2 u_1(x_1) u_2(x_2),$$

where w, w_1, and w_2 are called scaling constants in this situation, and $u_1(x_1)$ and $u_2(x_2)$ are called individual attribute utility functions. Since the function is scaled, we will have:

$u_1(3.5) = 0$, $u_1(0) = 1$, $u_2(24) = 0$, and $u_2(0) = 1$.

It can be shown that $w = (1 - w_1 - w_2)/w_1w_2$ (see pages 244–245 of Evans (2017)). Let's assume that $w_1 = .2$ and $w_2 = .5$; then w will equal 3. Hence the utility function can be written as:

$u(x_1, x_2) = .2u_1(x_1) + .5u_2(x_2) + 3(.2)(.5)u_1(x_1)u_2(x_2) = .2u_1(x_1) + .5u_2(x_2) + .3u_1(x_1)u_2(x_2)$

We have the function values for u_1 and u_2 for the best and worst values of X_1 and X_2, respectively, but not the functions themselves. Let us suppose that these functions are given by:

$u_1(x_1) = -.2857x_1 + 1.$ and
$u_2(x_2) = -.04167x_2 + 1.$

Note that both individual attribute utility functions are linear with a negative slope. The negative slope implies that these utility functions correspond to attributes that we want to minimize, as opposed to maximize.

We want to choose the decision that maximizes expected utility, or, in other words, that maximizes the expected value of a function (the utility function). For the first decision under consideration, use the court-appointed attorney, the outcome will be:

$X_1 = 0$, $X_2 = 0$, with probability of .3,
$X_1 = 0$, $X_2 = 24$, with probability = .7.

Substituting these attribute values into the utility function, we obtain:

$u(0, 0) = 1$ with probability of .3 and
$u(0, 24) = .2$ with probability of .7.

So the expected utility for this decision of using the court-appointed attorney is:

$.3 (u(0, 0)) + .7 (u(0, 24)) = .3(1) + .7(.2) = .44.$

If the defendant hires the attorney, the outcome will be:

$X_1 = 3.5$, $X_2 = 0$, with probability of .6,
$X_1 = 3.5$, $X_2 = 12$, with probability of .2, and
$X_1 = 3.5$, $X_2 = 24$, with probability of .2.

Substituting these attribute values into the utility function, we obtain:

$u(3.5, 0) = .5$ with probability of .6,
$u(3.5, 12) = .25$ with probability of .2, and
$u(3.5, 24) = 0$. With probability of .2.

So the expected utility of the second alternative, hire the attorney, is given by:

$.6(.5) + .2(.25) + .2(0) = .35.$

Comparing the expected utilities of .44 and .35, the alternative of using the court-appointed attorney gives the larger expected utility and is therefore the preferred alternative.

The main purpose of this example is to illustrate the computation of expected utility. The defendant in this case probably could have made the correct choice just by looking at the probabilistic outcomes, that is, without the use of a utility function. In Chapter 3, we include a much more involved example where maximization of expected utility is used to determine the pre-trial disposition of a defendant.

It is important to keep in mind that both multiattribute value functions and multiattribute utility functions are personal in nature; that is, for a group of people faced with a specific decision situation, either deterministic or probabilistic in nature, every one of them will have either a value function or utility function that is at least slightly different from everyone else in the group. Note that this does not necessarily mean that they would not rank the alternatives in the same way. For example, with one decision maker, the expected utilities for three alternatives might be given as alternative one: .47, alternative two: .68, alternative three: .55, while for a second decision maker, the expected utilities might be given as: alternative one: ..64, alternative two: .78, alternative three: .68. In this case the two decision makers would rank these alternatives in the same way: alternative one: third; alternative two: first; and alternative three: second.

2.3.5 Simulation Modeling and Analysis

Simulation refers to the broad collection of methods and applications to mimic the behavior of real systems, usually on a computer with appropriate software (page 1, Kelton, Sadowski, and Zupick (2015)). As such simulation allows one to re-create the operation of a complex system in a digital fashion on a computer. It is an especially useful technique when that system operation involves dynamic and probabilistic behavior over time.

With a simulation model of a system, one can experiment with different policies, procedures, and operations (via input to the simulation model) and examine the outputs from the model to determine the best policy or rank the various policies. The use of statistical analyses is important in the examination of the output because of the probabilistic nature of the simulation (which represents the probabilistic nature of the system). For example, in Chapter 3, we present a simulation model to examine various policies for pretrial conditions imposed by a judge on a defendant. Inputs to the model include the probability that the defendant will show for the trial, and the probability that the defendant will commit additional crimes while free on bail prior to the trial. This probabilistic representation of the defendant's behavior requires multiple (e.g., 10s or 100s) replications (runs) of the model to perform a proper statistical analysis. In the first replication the defendant may show for the trial and in the second replication the defendant may not show, according to the appropriate input probabilities.

One of the advantages of a simulation model in policy analysis is the fact that many different policies can be evaluated (and ranked in order of preference), typically in just a few minutes of time. Each replication of the model in these cases may represent several months or even years of simulated time. Hence, experimenting with the actual system in these cases would be impossible, not only from the consideration of the time involved, but also from the consideration of disruption to the actual system.

Another advantage of simulation as a tool for policy analysis is the fact that these models can output estimates for attribute values, and corresponding estimated values for multiattribute value functions or multiattribute utility functions, for any specific policy. Using multiple replications of the model, an estimate of the expected utility of a policy can be made when expected utility is the criterion of interest.

Finally, when a policy can be defined using values assigned to several control variables, there may be too many policies to evaluate using the model, given the combinatorial nature of the inputs. An example of this situation would occur when one has six different alternative programs of rehabilitation to implement with ten different categories of prisoners, as defined by age, sex, crime background, etc. In this situation, an alternative would be defined by the assignment of a rehabilitation program to a prisoner category. Hence, the number of combinations (or the number of alternatives) would be given by 6^{10}, more than 60 million alternative combinations. In this situation one may want to use an optimization procedure to implicitly enumerate many of the policies. Most simulation software packages include an optimization package which allows the user to interface their simulation model with an optimization process in order to find an estimated optimal policy.

Although a simulation model can be constructed using a general-purpose programming language, most such models are built with a simulation software package. Examples of these packages include Simio®, Arena®, ProModel®, AnyLogic®, SIMUL8®, and Flexsim®. Many of these packages employ a process-oriented approach in which the simulation represents a process, composed of various activities, through which entities flow. The activities can be performed by various resources, and are represented by icons which the model builder places on the computer screen.

For example, if one used Figure 1.1 as a guide to build a simulation model of prosecution, pretrial services, and adjudication for a criminal justice system, the entities would be the defendants flowing through the process. These entities would have descriptors called attributes. (Note that the word attribute has two different meanings in this book: performance measures in reference to a multiattribute value/utility function, and descriptors of entities in reference to a simulation model.) The attributes for the entities would be things like: crime accused of, demographic information, etc.

2.4 A REVIEW OF SELECTED APPLICATIONS OF ANALYTICS TO CRIMINAL JUSTICE SYSTEMS

In this section we will review some selected applications of analytics for the design and operation of criminal justice systems that are published in the literature. As compared to other systems, for example in the areas of health care, production, transportation, supply chains, etc., the applications in criminal justice have been relatively sparse. In addition, the applications involving multiple objective methods in criminal justice have been rare.

Much of the research into criminal justice systems involves studying the effects of policies, laws, and procedures, or studying and predicting criminal behavior. This is as opposed to optimization over an entire set of policies, laws, and procedures. Because of space limitations, the review provided here will not be exhaustive.

There are many journals and books which publish research addressing the area of criminal justice. The list of journals include the following: *Criminology and Public Policy, Journal of Drug Issues, Journal of Criminal Justice, American Journal of Criminal Justice, Journal of Experimental Criminology, Journal of Criminal Law and Criminology, The Prison Journal, Crime & Delinquency, Journal of Research in Crime & Delinquency, Journal of Quantitative Criminology, Security Journal, Criminal Justice Studies, Criminology, Punishment & Society, Criminal Justice and Behavior, Policing: An International Journal of Police Strategies & Management, Justice Quarterly, Journal of Drug Issues, Federal Probation*, and *Crime Science*.

There are also many University law journals available, such as the *Harvard Law Review* and the *Emory Law Journal*.

Science, PLOS ONE, Kybernetes, and *Psychiatric Services* are also journals which sometimes contain research in criminal justice.

Finally, journals/proceedings which contain publications in operations research and systems, sometimes with papers that address criminal justice include *Operations Research; Management Science; INFORMS Journal on Applied Analytics* (formerly called *Interfaces); Decision Analysis; Omega; Journal of the Operational Research Society; European Journal of Operational Research; Socio-Economic Planning Sciences; IEEE Transactions on Systems, Man, and Cybernetics;* and *Proceedings of the Winter Simulation Conference*. The *INFORMS Journal on Applied Analytics* is of note because of its focus on applications as opposed to theory.

Magazines or media organizations of a more general nature which sometimes contain articles on subjects related to criminal justice include *The Atlantic, Politico, The Conversation,* and *Mother Jones*. With respect to at least some of these publications, the reader should be aware of a rare bias which can affect the argument made.

Some of the earliest research in the application of OR techniques to the criminal justice system involved the use of simulation to model the entire system, much like the system illustrated in Figure 1.1. An example of a publication representing this research is Blumstein and Larson (1969). Their model represents flows of criminals/defendants from one stage to another of a criminal justice system. One of the features of the model is that it represents the flow of some of the prisoners/defendants back into the system after they have left it (a feedback loop). The model is illustrated using data collected in the state of California through 1965. The outputs from the model included projections of workloads, costs, and manpower requirements.

The research of Blumstein and Larson and additional early research of a similar type is described in Chaiken et al (1976) and Bohigian (1977). In particular, Bohigian (1977) divides the models reviewed in his paper according to the categories of criminal justice planning models; police patrol models; prosecution models; forensic science models; court operation models; juror management models; corrections, parole, and probation models; juvenile delinquency models; police communication models; and offender tracking models.

Auerhahn (2002) investigated the effects of sentencing reforms (for example, the three-strikes law) in the state of California on the percent of the prison population that is elderly. In a later work, Auerhahn (2004) used simulation to investigate the influence of California's Substance Abuse and Crime Prevention Act of 2000 on California's prison population to the year 2020.

Berenji, Chou, and D'Orsogna (2014) used simulation to compare in a general sense rehabilitation programming vs. prison as a policy for nonviolent drug offenders on recidivism.

In an application like that found in many systems which involve the use of processes to accomplish goals, Larson, Cahn, and Shell (1993) used Monte Carlo simulation to improve the arrest-to-arraignment process in New York City. Using the simulation, they developed policies that allowed for the average time for arrest-to-arraignment to be reduced from about 40 hours to 24 hours, at an annual savings of $3.5 million.

Most of the applications of simulation mentioned above involve the use of **continuous simulation** in which the continuous flows of prisoners/defendants from one state (e.g., convicted, paroled, or free on bond) to another are represented. In a different simulation-oriented approach, Merlone, Manassero, and Raffaello (2016) employed agent-based simulation to model the behavior of 200 individual criminals.

A important aspect of simulation is the concept of validation, that is, whether the simulation is an accurate representation of the system being modeled. Berk (2008) provides some guidance in this area for simulation models of criminal justice systems.

Analytical models such as those involving queueing theory and other types of probabilistic representations have also been used to address issues in criminal justice systems. For example, consider the fact that states have different requirements regarding jury size. Nagel (1981) addressed the decision problem of jury size using probabilistic modeling; the issues addressed included the facts that (1) if there are too many jurors, then many guilty defendants will be declared not guilty (called a type I error) and (2) if there are too few jurors, then some innocent defendants will be declared guilty (called a type II error). Gorski (2012) notes that the US Supreme Court has been making intuitive decisions regarding jury size, instead of investigating the situation on the basis of mathematics.

Harris and Thlagarajan (1975) used queueing theory to aid decision makers in the District of Columbia to manage the capacities of their halfway houses. Yablon (1991) also used queueing theory to model prison populations as an aid in capacity planning for prisons.

Caulkins (1993) used some sophisticated mathematical modeling to show that, counter to intuition, a zero-tolerance policy with respect to usage of illegal drugs can actually result in an increase in usage of these drugs. (A similar situation is mentioned in Chapter 4 where it is suggested that adding more police could actually result in more crime.)

Another popular methodology in operations research is optimization. However, as noted by Blumstein (2002), there have been relatively few applications of this methodology to criminal justice systems because of the complex criteria associated with these systems.

This section has discussed only a few of the applications of operations research methodology to criminal justice systems in order to give the reader a flavor for this area. Additional detail can be found in other review papers such as Maltz (1994); Barnett, Caulkins, and Maltz (2000); Blumstein (2007); and Bae and Evans (2019).

USING MULTIPLE OBJECTIVE ANALYTICS TO ADDRESS ISSUES ASSOCIATED WITH CRIMINAL JUSTICE SYSTEMS

3.1 INTRODUCTION

Chapter 3 addresses two of the major issues of criminal justice systems in recent years: privatization of prisons and bail reform.

Following the introduction, the second section of the chapter addresses the issue of privatization of prisons. In particular, the section focuses on the decision of whether prisons should be public or private. As part of the discussion, data on prisons and the motivation for privatization is addressed. The advantages and the disadvantages of privatization are presented, along with the positions of some prominent Democrats and Republicans on the issue. A discussion of how the two major political parties have recently addressed the issue is given. Finally, an illustrative example involving the assessment of a multi-attribute value function for the scoring and ranking of proposals for privatization from private companies is presented. The function allows for (1) the representation of several quality measures for a prison, and (2) the modeling of the trade-offs between cost and quality for a prison.

The third section of the chapter discusses bail reform. In particular, the decision problem of the choice of pretrial conditions for a defendant is considered. The bail process and its associated difficulties are described, along with recent trends in bail reform. An illustrative example involving a Monte Carlo simulation model of a bail decision process is presented. The model represents various decisions which a judge could make regarding pretrial conditions for a defendant, and accounts for various costs and probabilities. The criterion used for the ranking is maximization of expected utility, with attributes of concern for both the defendant and the public.

3.2 PRIVATIZATION OF PRISONS

3.2.1 Data on Prisons and the Motivation for Privatization

From 1980 to 2000, the incarceration rate in state and federal prisons in the United States more than tripled, from 139 per 100,000 population to 478 per 100,000 population (Mumford, Schanzenbach, and Nunn, 2016). This tremendous increase was the result of more strict laws/acts such as President Reagan's Anti-Drug Abuse Act of 1986, and the Violent Crime and Law Enforcement Act of 1994 (Clinton Crime Bill) and its Three-Strikes Provision (Violent Crime and Law Enforcement Act of 1994, 1994), among other things. Pfaff (2016) noted that this increase was due more to higher rates of admission than to longer sentences. As a result of this increase in incarceration rate, state and federal prisons became increasingly overcrowded during this period.

Many states attempted to increase their prison capacity during this period. However, even though much of the public was interested in "getting tough on crime", the cost associated with doing so was often rejected. For example, during the 1980s voters turned down about 60% of prison bond referenda (Mumford, Schanzenbach, and Nunn, 2016).

In addition to the overcrowding, the costs of operating the prisons became exceedingly high. One attempt to reduce cost was the introduction of the concept of privatization. Even though the first private jail in the United States was at San Quentin in 1852 (Private Jails in the United States, 2017), this more recent surge in privatization was started at the state level in St. Mary, Kentucky, in 1986 at the Marion Adjustment Center. The first contract to a private company at the federal level was instituted in 1997 (Mumford, Schanzenbach, and Nunn, 2016).

While the focus on this section of the book is on issues associated with private companies in the building and operation of entire prisons, private entities have also been employed in providing individual services for prisons, such as food, health, transportation, and cleaning services, among others.

In 2017, private prisons (including state and federal) in the United States housed 121,718 people, about 8.2% of the total prison population. Since 2000, the number of prisoners incarcerated in private facilities has increased by 39%; however, there has been a decline in more recent years, as the private prison population reached its peak of 137,220 in 2012 (Private Prisons in the United States, 2019). Pauley (2016) provides a year-by-year history of the various events associated with the privatization of prisons.

3.2.2 Advantages and Disadvantages of Privatization

With respect to privatization, there are several supposed advantages and several supposed disadvantages, as well as disagreements about these advantages and disadvantages. Two of the (disputed) advantages include the following: (1) at the state level, states can have prisons built by private companies without the approval of voters through the use of bond referenda and (2) the use of private organizations for building and operating prisons can promote competition (and hopefully lower costs and improved quality). Three of the (disputed) disadvantages of privatization are: (1) there can potentially be a lack of transparency in how the prison is run, (2) cost (to the private entity) can become an overriding factor in the prison's operation (Bryant, 2020), and (3) since typically a private company will make more profit with more prisoners, they may have a tendency to lobby for stricter laws and longer sentences.

The two largest private prison companies are CoreCivic (formerly Corrections Corp) and GEO Group (formerly Wackenhut Corrections Corporation). These companies are aware of some of the poor perceptions associated with them in both state and federal governments. Hence, their websites (https://www.corecivic.com/ for CoreCivic and https://www.geogroup.com/ for GEO Group, both accessed on May 1, 2020) address the importance of community engagement, non-involvement in government decision making with respect to criminal justice, quality programs, and policies to reduce recidivism. A portion of the Geo Group web page addresses the importance of not engaging in political activity in criminal justice:

> *GEO does not take a position on or advocate for or against criminal justice and immigration policy related to criminalizing certain behaviors, determining the length of criminal sentences, or immigration enforcement policies.* (https://www.geogroup.com/Political _Engagement, accessed on May 1, 2020)

This policy as noted in the above quote tends to contradict some of the past actions by these private companies. For example, CoreCivic and GEO Group have both opposed several bills in the past similar to the 2005 bill sponsored by Senator Ted Strickland involving the requirement that private prisons holding federal prisoners comply with the Freedom of Information Act (Pauley, 2016).

3.2.3 Democrats vs. Republicans with Respect to Privatization

There is certainly much disagreement between Democrats and Republicans with respect to private prisons. Much of the disagreement stems from the evaluation on the quality side of the issue, with respect to medical care, and most specifically with respect to medical care at immigration detention centers. For example, during the campaign for the 2016 Democratic presidential nomination, Bernie Sanders pledged to ban contracts to private entities for the operation of all federal, state, and local prisons. Hillary Clinton pledged to do something similar for immigration detention centers (Hing, 2016). It should also be noted here that in addition to the immigration detention centers, a large percentage of the incarcerated held in private prisons in general are undocumented immigrants (Gaes, 2019).

During the campaign for the 2020 Democratic presidential nomination, Elizabeth Warren proposed the elimination of the use of private companies at the federal level, while several of her competitors proposed restrictions on private companies. On the other hand, during the first two and one-half years of Donald Trump's presidency, the United States Immigration and Customs Enforcement (ICE) awarded $480 million and $331 million, respectively, to the two private companies mentioned above for the operation of immigration detention centers (Layne, 2019).

3.2.4 The Literature on Private vs. Public Prisons

Articles from the literature comparing public prisons and private prisons provide an overall picture that is inconclusive. The comparisons are often difficult because (1) private prisons often have restrictions on who they accept as inmates, (2) operational aspects of private prisons are often not transparent, (3) studies often do not control for factors such as capacity and age of the facility (Perrone and Pratt, 2003), and (4) prisoners can often spend time in both public and private facilities (Spivak and Sharp, 2008) during their incarceration.

Each of these published articles could be placed into one of three categories: studies involving the comparison of public and private prisons, reviews of these studies, and reviews of the reviews. For example, with respect to the second category (reviews), there are the articles by Pratt and Maahs (1999), Segal and Moore (2002), Volokh (2002), Perrone and Pratt (2003), Volokh (2013), and Pratt (2019).

A recent study of El Sayed et al (2020) concluded that there was little difference between the rates of misconduct in public and private prisons in Texas, except for the fact that prisoners in private facilities were 10% less likely to engage in **severe** violence as compared to those in public facilities.

In a study done in the state of Florida, Lanza-Kaduce et al (1999) determined that private prisons had lower rates of recidivism than public prisons.

Pratt and Maahs (1999) analyzed 33 cost-effectiveness evaluations of public and private prisons from 24 independent studies. They determined that there was not much difference in cost-effectiveness between public and private prisons, and that facility characteristics such as age, economy of scale, and security level were the strongest predictors of a prison's cost.

Segal and Moore (2002) reviewed 28 studies, including many of the same ones as Perrone and Pratt (2003). They determined that 22 of the 28 studies concluded that there were significant cost savings associated with private prisons; these private prisons were judged to be 3.5%–17% less costly than public prisons. They also concluded that private prisons provide at least the same level of quality as public prisons. As noted by Perrone and Pratt the analysis on the quality side of performance by Segal and Moore was not as rigorous as that on the cost side.

Volokh (2002) concluded that private prisons outperformed public prisons on both cost and quality dimensions. In a later publication, discussed below, Volokh (2013) modified his conclusion on the quality dimension.

Perrone and Pratt (2003) reviewed nine studies on seven different performance measures or domains, for a total of 63 combinations. They determined that private prisons surpassed public prisons on 17 of the comparisons, public prisons surpassed private prisons on 6 of the comparisons, while the remaining 40 comparisons were inconclusive with respect to producing a winner. It is difficult to make an evaluation here because, obviously, some of the 63 combinations would be more important than others.

Volokh (2013) determined that there was no strong evidence to determine that either side, private or public, performed better than the other with respect to the quality measure. One of the important features of Volokh's work was a detailed breakdown of the various sub-measures of quality; another important attribute of Volokh's paper was a discussion of how a private organization could manipulate (or game) some of these sub-measures so that their prisons would appear to perform better than they actually did. For example, a prison can have prisoners released in areas where policing is weak, thereby resulting in better measures of recidivism.

Pratt (2019) reviewed the literature on cost–benefit analysis with respect to privatized corrections. He noted the limited role that this type of analysis has on policy discussions.

One of the most comprehensive reviews of the literature in the comparison of public and private prisons is provided by Gaes (2019). Some of his major conclusions include the following: (1) the costs and quality of privately operated prisons are about equal to that for publicly operated prisons, although evidence for this conclusion is not strong, owing at least in part to the lack of good studies in the area; (2) private facilities seem to have a record worse than public facilities when it comes to recidivism; (3) there is no corroboration that privatization of prisons was a cause for the growth in incarceration in general that crested in 2009; and (4) the relevant jurisdiction needs to closely monitor private prisons, especially for those actions, programs, and policies corresponding to the quality of the service provided. The types of services that should be monitored include health care, safety/security, food services, educational services, and rehabilitation programs.

Gaes focused on two primary dimensions in his work: cost and quality. For the primary justification for the first conclusion listed in the previous paragraph, he studied five reviews from the literature. In three of these reviews, the authors concluded that there were no basic differences in either cost or quality between public and private prisons, while in two of the reviews, the authors concluded that private prisons operate at a much lower cost, but at similar quality as public prisons.

Although it was thought at one time that the advent of competition from privatization of the prison industry would result in lower costs in the public sector, there is little evidence of this actually occurring (Burkhardt, 2017 and Blumstein, Cohen, and

Seth, 2007). This lack of competition is indicated by the fact that most prisoners incarcerated in private prisons are incarcerated in prisons operated by one of only two companies mentioned earlier: CoreCivic and the Geo Group.

The evidence so far suggests that private prisons are not appreciably better or worse than public prisons. Moreover, valid studies of the subject area are difficult to accomplish for several reasons:

- Controlling for type of prisoner is extremely difficult, especially since prisoners can be moved from one facility to another, and private prisons can often decide whether to accept certain types of prisoners,
- Whereas cost is difficult to measure in and of itself, the measurement of quality is even more so because of its various dimensions, many of which are subjective in nature; in fact, Volokh (2013) notes that "good quality measures are rarely used" in the study of prisons. In addition, to provide an overall rating to a prison, one must consider the trade-offs between cost and quality.
- Many studies do not address recidivism since it occurs after the fact of imprisonment. Recidivism could easily be thought of as the most important measure of "quality"; however, it should be thought of as an output measure. Instead, the various studies would tend to measure inputs for quality, such as the number of prisoners to be involved in rehabilitation programs.

For additional literature involving the comparison of public and private prisons, see Lombardo (2014) and Lundahl et al (2009).

In the following illustrative example we address the second issue discussed above, the difficult nature of the measurement of quality and the trade-offs between cost and quality through the hypothetical assessment of a multiattribute value function over attributes which represent the various measures of cost and quality.

Such a multiattribute value function could be employed in either or both of at least two ways:

(1) in the evaluation of proposals submitted by private companies, such as CoreCivic and the GEO group, for the development and operation of federal or state prisons or (2) in the comparison of private and public prisons. An example of the type of information that would be required from a private company from a state such as Florida is given in Florida State Bill 238 (FL S.B. 238, 2012).

3.2.5 An Illustrative Example: Development of a Multiattribute Value Function to Compare Prisons

As noted in Chapter 2, a multiattribute value (MAV) function is one which maps a group of attribute values into a real number. These attributes (independent variables for the function) can be thought of as important measures for the evaluation and ranking of alternatives. Normally the function is scaled so that it will attain a value between 0 and 1. An alternative will be preferred over another alternative if the first alternative attains a higher value on the MAV function than the second alternative. If the two alternatives attain the same value, then they are equally preferred.

In the illustrative example described in this section, we assume a hypothetical evaluation and ranking situation like the following: a state or federal government wants to rank several proposals for the construction and operation of a private prison. The companies submitting the proposals have designed these proposals to correspond to a format which provides data according to the attributes described in this section. In other words, the companies would be responding to a request-for-proposal which would require the relevant data.

A second use of the MAV function described here could involve a comparison of existing prisons, some private and some public. It would be expected that the prisons evaluated would be similar in terms of size, types of prisoners, and facility age to make the comparison fair. In performing this exercise, one would supposedly be able to come to some conclusions about private versus public prisons.

One of the primary characteristics of the MAV function developed in this section is that it allows for trade-offs between cost and quality. As noted by Volokh (2013), very few studies involving the comparison of public to private prisons have involved the evaluation of cost and quality simultaneously.

3.2.5.1 Determination of Objectives and Attributes

The first step in forming a MAV function for a decision situation is to determine the attributes (independent variables) over which the function will be formed. These attributes can be thought of as measures for the objectives, which can be formed as a hierarchy, with the more general objectives toward the top of the hierarchy (e.g., optimize prison quality) and the more specific objectives (e.g., maximize percentage of prisoners enrolled in drug rehabilitation programs) towards the bottom. An attribute can be associated with each objective at the lowest level of the hierarchy.

As noted in Chapter 2, these attributes can be either natural or constructed in nature. Typically, constructed attributes (e.g., prison quality measured on a scale from 0 to 5) would be employed as measures for general objectives toward the top of the hierarchy, while natural attributes would be used as measures for objectives toward the bottom of the hierarchy (e.g., percentage of inmates enrolled in drug rehabilitation programs).

In addition to the classification of attributes as being either natural or constructed, they can also be classified as being input measures or output measures. An example of an input measure would be the number of dollars spent per prisoner per day on educational/rehabilitation programs. An example of an output measure would be the percent of prisoners who are re-arrested within one year of being released from prison. There are both advantages and disadvantages to the use of input measures vis-à-vis output measures. Input measures such as the number of dollars spent on prison rehabilitation programs are relatively easy to measure and control (through policy); however, they may not directly measure the objective for which we are interested (e.g., reduce recidivism) as, for example, the percentage of released inmates who are re-arrested within a year of release. On the other hand, it is very difficult to determine an accurate attribute for an output objective such as "optimize reduction in recidivism"; the difficulty here is that released prisoners who are considered in the population for recidivism may have only spent a short amount of time in the prison situation being measured.

As noted above, input measures/attributes may also often be proxy attributes—that is attributes which do not directly measure an objective; see pages 41–44 of Evans (2017). For example, the amount of funds spent per prisoner on education/rehabilitation programs does not directly measure an objective like "minimize recidivism".

There are many different, but often complementary, approaches for developing a hierarchy of objectives and associated attributes, as described in Chapter 2.

Probably the first set of performance measures developed to evaluate prisons was the Correctional Institutions Environment Scale, developed originally by Moos (1987). The measures, including involvement, support, autonomy, etc., for this scale are nebulous in nature though. As described by Volokh (2013), others who have developed sets of measures include Burt (1981) and Logan (1993), from whom we borrow for the discussion that follows.

We will employ the method of specification/means-ends as discussed in Chapter 2 to develop our hierarchy of objectives and attributes. With this method, an analyst questions a subject-matter expert (or experts) about a situation. To expand from an objective at a higher-level to the next lower-level objectives, the subject-matter expert is asked the basic question of "What means do you use to achieve the higher-level objective?". Alternatively, the subject-matter expert could be requested to: "Specify what you mean by this higher-level objective".

At the top level of our hierarchy, we will have the objective of optimize the choice of the best prison/proposal for a prison. The question asked at this point would be: "What means would you use to achieve this choice of best prison/proposal?". The answer would be to choose the prison/proposal that allows (1) the minimization of cost and (2) the maximization of quality. (Note that in developing the hierarchy, we are not concerned about trade-offs between the objectives—that comes later, through the development of the MAV function.)

With respect to the objective of minimization of cost, one could expand this further into the different categories of cost, for example, personnel cost, facility cost, etc. However, a typical attribute used in this area is "cost per prisoner per day". So, for the cost side of the hierarchy, we will stop with the single objective of minimization of cost, with the associated attribute of cost per prisoner per day. One must be careful in deciding what goes into the cost per prisoner per day, especially since we will be trading this cost off against various measures of quality. For example, the state of Texas came up with an initial cost estimate of $27.62 per prisoner per day in 1987 as an initial basis for pay to private organizations for the operation of prisons. After considering things like building and program costs, however, the basic analysis resulted in raising the cost from $27.62 per day to $41.67 per prisoner per day (page 350 of Volokh (2013)).

With respect to the objective of maximization of quality, several sub-objectives could be specified. The sub-objectives specified here were obtained through a perusal of the literature, including Logan (1993). These sub-objectives for maximization of quality are: (1) optimize security, (2) optimize safety, (3) optimize order, (4) optimize health, both mental and physical, of the prisoners, (5) optimize rehabilitative and educational programs, (6) optimize living conditions, and (7) optimize management efficiency.

Security refers to both keeping prisoners from escape as well as keeping contraband material (e.g., drugs and weapons) from entering the prison and from transfer between areas of the prison. Safety refers to keeping staff and inmates safe from harm resulting from both assault and other hazards such as food/environmental hazards. Order refers to minimization of inmate misconduct and disturbances. Health refers to having proper procedures, personnel, and facilities to handle inmate illness, especially when contagion is a possibility; mental health is also a consideration here. Rehabilitative and educational programs refer to things like classes for drug and alcohol rehabilitation, and educational opportunities that will be helpful for job opportunities following release from prison. Living conditions entail things like quality of food, sanitation, visitation policies, cleanliness, and recreational facilities and opportunities among other things. Management efficiency involves things like policies for handing staff grievances, management structure, staff training programs, staff turnover rate, etc.

Each of these seven main sub-objectives could be further subdivided. For example, the second sub-objective, optimize safety, could be separated into the two additional sub-objectives of optimize safety for staff and optimize safety of inmates. However, there is a tradeoff involved here in that further subdivision (and correspondingly a greater number of total attributes) while leading to increased accuracy in terms of representing the preferences of the decision maker, also requires much more effort in the overall analysis, for example when we come to the assessment of the MAV function.

The attribute for the objective of minimization of cost has already been determined as cost per prisoner per day. Let us suppose that the minimum value (best value from the standpoint of the government that will be paying this) for this attribute is $80, and the maximum value (worst value) is $110. As mentioned earlier, this attribute is a natural attribute, as opposed to a constructed attribute.

Each of the seven quality sub-objectives contains many facets; hence, determining a natural attribute for each of these sub-objectives would be extremely difficult. Instead, we will use a constructed attribute to measure each of these sub-objectives. These constructed attributes, called the caliber of the associated objective achievement, will each be associated with a scale having a worst possible value of 1 and a best possible value of 5, and appropriate intermediate values of 2, 3, and 4. Associated with each of the numbers on a scale would be a description of what would be expected of a situation involving an attribute that would be assigned that value. The cost (or the amount spent) associated with achieving a specific value for the attribute is not a consideration in determining the caliber/value of the attribute, as measured on the scale from 1 to 5.

A reason for not considering the cost associated with a quality area of concern is the fact that this cost is supposed to be reflected in the first attribute, cost per prisoner per day. In addition, the amount of money spent in an area does not necessarily reflect how well the associated objective is achieved. For example, one situation may involve spending more on staffing than another situation, but because the allocation of staff both by time and location is much better in the second situation, that situation will achieve a better score on the attribute of "caliber of security".

Associated with each number (5, 4, 3, 2, and 1) on the subjective scale for an attribute should be a description of what is meant by an alternative which is given that score. For example, for the objective of "optimize rehabilitative and educational programs", we might have something like:

5: All eligible prisoners who desire it will have access to educational programs and rehabilitative programs; prisoners are given a variety of programs from which to choose; instructors of the programs are highly qualified with appropriate education and experience.

4: Most, but not all, prisoners who desire it will have immediate access to educational and rehabilitative programs; a few of the instructors will have only a minimum of experience; etc.

Additional descriptions could be given for ratings of 3, 2, and 1.

We use the following notations for the attributes:

X_1 represents the cost per prisoner per day, from its worst possible value of \$110 to its best possible value of \$80,

X_2 represents the caliber of security on a subjective scale of 1 (worst caliber) to 5 (best caliber),

X_3 represents the caliber of safety on a subjective scale of 1 (worst caliber) to 5 (best caliber),

X_4 represents the caliber of order on a subjective scale of 1 (worst caliber) to 5 (best caliber),

X_5 represents the caliber of health on a subjective scale of 1 (worst caliber) to 5 (best caliber),

X_6 represents the caliber of programs on a subjective scale of 1 (worst caliber) to 5 (best caliber),

X_7 represents the caliber of living conditions on a subjective scale of 1 (worst caliber) to 5 (best caliber),

X_8 represents the caliber of management efficiency on a subjective scale of 1 (worst caliber) to 5 (best caliber).

Note that X_i represents the relevant attribute i for i = 1, ..., 8 and x_i represents a value for that attribute. For example, if we want to refer to a cost per day for a prisoner of $90, we will write x_1 = $90. If we want to refer to the cost per day for a prisoner without specifying a value, we will just write X_1. In addition, we let x_i^b denote the best value for attribute i (which is $80 for attribute X_1 and 5 for all other attributes) and x_i^w denote the worst value for attribute i (which is $110 for attribute X_1 and 1 for all other attributes).

The best and worst values for any attribute would be the best and worst values that would be attained over all alternatives under consideration. Hence, in some cases, the difference in the score between 1 and 5 might be very little because all alternatives would score approximately the same on that attribute.

3.2.5.2 Assessment of the MAV Function

Once the attributes and the associated worst and best possible values for these attributes have been determined, the assessment process for an MAV function can start. As with the process to determine the hierarchy of objectives and attributes, this assessment process involves interaction between analysts and decision makers.

The typical first step is to determine the form of the MAV function. We will assume that this function is a scaled, additive function (see pages 92–112 of Evans (2017)) as given by:

$$v\left(x_1, x_2, x_3, \ldots, x_8\right) = \sum_{i=1}^{8} c_i v_i\left(x_i\right),$$

where v_i are individual attribute value functions such that $v_i\left(x_i^b\right) = 1$ and $v_i\left(x_i^w\right) = 0$ for i = 1, ..., 8; and c_i are scaling constants such that $\sum_{i=1}^{8} c_i = 1$. Such an additive MAV function is appropriate when the decision maker's preference structure meets certain conditions, which are usually verified through answers to questions by the decision maker. Even if the decision maker's preference structure does not meet the requirements exactly for an additive function, such a function may very well be a good approximation for the situation.

There are various approaches for determining the c_i values and the v_i functions that will completely specify the MAV function (see 94–111 of Evans (2017)). Here we use an approach called the simple multiattribute rating (SMART) technique (see Edwards (1977), Edwards and Barron (1994), and pages 278–286 of von Winterfeldt and Edwards (1986)).

The first step in the SMART approach is to specify the individual v_i functions. This is accomplished by setting $v_i(x_i^b) = 100$ and $v_i(x_i^w) = 0$ for $i = 1, ..., 8$. Note that this approach amounts to a rescaling of the individual attribute value functions from values of 0–1, to values of 0–100. For most decision makers, thinking of values on a scale of 0–100 is easier than thinking of values on a scale of 0–1. When the assessment process is finished, we can rescale the individual attribute functions.

Following the setting of these two initial values, intermediate values (between 0 and 100) are set for the individual attribute value functions by the decision maker. The key here is for the decision maker to think in terms of the differences in values for the attribute between the two extremes. For example, suppose for the cost attribute X_1, a cost of $100 per day (or more) per prisoner will cause a great deal of difficulty from a political standpoint. In that case, the direct ratings for the individual attribute value function might be set as:

$$v_1(80) = 100, v_1(85) = 90, v_1(90) = 80, v_1(95) = 65, v_1(100) = 35,$$
$$v_1(105) = 18, \text{ and } v_1(110) = 0.$$

Note that the values associated with costs that are $100 or greater are much less than they would be if v_1 were a linear function.

In a similar fashion, we can determine values for the other individual attribute value functions:

$$v_2(5) = 100, v_2(4) = 82, v_2(3) = 60, v_2(2) = 32, v_2(1) = 0,$$
$$v_3(5) = 100, v_3(4) = 75, v_3(3) = 50, v_3(2) = 25, v_3(1) = 0,$$
$$v_4(5) = 100, v_4(4) = 90, v_4(3) = 80, v_4(2) = 50, v_4(1) = 0,$$
$$v_5(5) = 100, v_5(4) = 80, v_5(3) = 55, v_5(2) = 26, v_5(1) = 0,$$
$$v_6(5) = 100, v_6(4) = 77, v_6(3) = 57, v_6(2) = 20, v_6(1) = 0,$$

$v_7(5) = 100$, $v_7(4) = 82$, $v_7(3) = 60$, $v_7(2) = 22$, $v_7(1) = 0$, and
$v_8(5) = 100$, $v_8(4) = 95$, $v_8(3) = 88$, $v_8(2) = 35$, $v_8(1) = 0$.

In answering the questions concerning the values associated with levels of an attribute, note that the decision maker does not consider the level of that attribute in comparison with the levels of other attributes. Comparisons between different attributes are considered through the determination of the weights for the MAV function.

As an example of how to interpret one of the individual attribute value functions, consider v_8 for X_8, management efficiency. The values shown for v_8 indicate that if management efficiency is at a level of 3 or above, the decision maker will be fairly satisfied with that attribute value.

For attribute values which are not equal to one of the assessed points above (e.g., $x_3 = 4.5$, or $x_1 = 88$) one can either employ a linear interpolation between two points, or one can use the value of a function fitted through all five (in the case of X2 through X8) or seven (in the case of X1) points. For this example, we will employ linear interpolation. For example, the value associated with a cost per prisoner per day of \$88 would be 84 (or .84 after rescaling from 0 to 1).

The next step of the SMART process would be to determine the weights for the individual attribute value functions. We will employ an approach called the swing weighting method (pages 106 and 107 of Evans (2017)) to determine the weights: c_i for $i = 1, \ldots, 8$. The first step of this approach is to ask the decision maker which attribute he or she would like to move from its worst value to its best value given that all other attributes are at reasonable values. This attribute is assigned a relative weight of 100. The decision maker is then asked to determine, given all other attributes are at reasonable values, which attribute would be ranked second for this operation; that attribute is assigned a relative weight less than 100. This procedure is continued until all attributes have been accounted for in this ranking procedure.

Let's denote the relative weights as r_i for $i = 1, \ldots, 8$, and let's suppose that the decision maker answered the questions alluded to in the previous paragraph in such a way that the attributes would be ranked in decreasing order of preference as:

X_1 (cost) with $r_1 = 100$, X_3 (safety) with $r_3 = 85$, X_2 (security) with $r_2 = 80$, X_5 (health) with $r_5 = 75$, X_6 (programs) with $r_6 = 60$, X_4 (order) with $r_4 = 55$, X_7 (living conditions) with $r_7 = 50$, and X_8 (management efficiency) with $r_8 = 40$.

These relative weights, r_i, are normalized to give the weights used in the MAV function as follows:

$$c_i = r_i / (r_1 + r_2 + r_3 + r_4 + r_5 + r_6 + r_7 + r_8).$$

Hence, we obtain the value for c_1 of:

$$c_1 = 100 / (100 + 85 + 80 + 75 + 60 + 55 + 50 + 40) = .18.$$

Computing the values for the other weights in an equivalent fashion, we obtain: $c_2 = .15$, $c_3 = .16$, $c_4 = .10$, $c_5 = .14$, $c_6 = .11$, $c_7 = .09$, and $c_8 = .07$.

3.2.5.3 Ranking Proposals from Private Companies

Let's suppose that there are a group of officials for a state government, charged with the evaluation of proposals from four different private firms for the operation of a state prison. These four firms have submitted proposals in response to a request for proposals, designed to correspond to the discussion above; that is, involving the attributes described. Based upon their readings of the proposals, the government officials give the following attribute values for the proposals.

> Proposal from Firm 1: $X_1 = \$101$, $X_2 = 3$, $X_3 = 5$, $X_4 = 5$, $X_5 = 5$, $X_6 = 4$, $X_7 = 3$, $X_8 = 3$,
> Proposal from Firm 2: $X_1 = \$96$, $X_2 = 5$, $X_3 = 2$, $X_4 = 4$, $X_5 = 1$, $X_6 = 5$, $X_7 = 2$, $X_8 = 5$,
> Proposal from Firm 3: $X_1 = \$80$, $X_2 = 2$, $X_3 = 3$, $X_4 = 4$, $X_5 = 3$, $X_6 = 4$, $X_7 = 5$, $X_8 = 2$,
> Proposal from Firm 4: $X_1 = \$110$, $X_2 = 1$, $X_3 = 2$, $X_4 = 4$, $X_5 = 1$, $X_6 = 3$, $X_7 = 2$, $X_8 = 1$.

From comparison of these attribute scores, one can determine that the proposal from Firm 4 is **dominated** (see pages 76–78 of Evans (2017)) by the proposals from each of the other firms, that is, each of the other firm's proposals performs at least as well on each attribute, and better on at least one attribute. Hence, the proposal from Firm 4 can be eliminated without further consideration. The other outcomes, from Firms 1, 2, and 3 are what are called the set of nondominated outcomes—that is not one of the set of scores from Firms 1, 2, or 3 is dominated by a set of scores from one of the other Firms.

Using the MAV function to score the proposal from Firm 1, we obtain the following MAV score:

MAV score for Firm 1: .18(.316) + .15(.6) + .16(1) + .1(1) + .14(1) + .11(.77) + .09(.6) + .07(.88) = .75.

In a similar fashion, we obtain scores for the other two firms as:

MAV score for Firm 2: .59, MAV score for Firm 3: .67.

Hence, we would rank firms' proposals in decreasing order of preference as Firm 1, Firm 3, Firm 2, Firm 4.

3.3 BAIL REFORM

3.3.1 The Bail Process

At a defendant's first appearance before a judge, called the arraignment, he or she is informed of the charges and is asked to enter a plea. At this point a decision is made concerning the disposition of the defendant until the start of the trial. The judge will either (1) remand the defendant to the local jail, which is typically done only in exceptional cases such as murder; (2) release the defendant on his/her own recognizance (OR); or (3) specify an amount for bail, and/or other conditions so that the defendant can be released until the start of the trial.

Under the third alternative in the previous paragraph, the judge has historically had broad discretion in deciding on the form of the bond: cash only, surety, or property. Surety means that a traditional bail bond is obtained from a professional bondsman by paying 10% of the total—which the professional bondsman keeps as his or her fee. (More about this later.)

The requirement of cash only (in an amount more than the defendant can likely afford) effectively gives the judge the power to impose "preventive detention". This could be done with a defendant who has multiple prior crimes, possibly because of drug abuse. A judge noting that pattern of behavior may eventually decide enough is enough and set bond in an amount intended to keep the person in jail.

If the judge sets a bail amount that the defendant cannot pay, or does not want to fully pay, that defendant usually has the option of paying 10% of the bail to a bail bond company, which will then pay the entire bail amount. When the defendant appears for his/her trial, the bail money is returned to the bail bond company; however, the defendant does not receive the 10% back. If the defendant paid the entire amount for the bail without the help of a bail bond company, the entire amount is returned to the defendant once the case is over. In case of conviction, additional fees may be deducted prior to the amount being returned to the victim.

The other conditions referred to in the third alternative involve things like house arrest with possibly the use of electronic monitoring devices, curfews, regular check-ins, alcohol/drug locking systems for automobiles, and voluntary drug testing and/or rehabilitation programs, among other things ("Bail Amounts by Crime," 2019).

3.3.2 Difficulties Associated with the Current Bail Process

The concept of bail dates to medieval England (Josephson, 2018). The eighth amendment of the United States Bill of Rights prohibits excessive bail. Specifically, this amendment states:

> *Excessive bail shall not be required, nor excessive fines imposed, nor cruel and unusual punishments inflicted.*

However, what constitutes **excessive bail** is not defined by the US Constitution. It is clear however that there are many defendants who are incarcerated in local jails because they cannot afford to pay the full amount of their bail, or even the 10% required of a bail bondsman. For example, of the 740,700 inmates incarcerated in local jails in the United States at mid-2016, 458,600 (or about 62%) had not been convicted of a crime (Zeng, 2018). Most of these 458,600 defendants were incarcerated because they could not afford to pay bail; however, there is no data on exactly how many incarcerated defendants fall into this category of not being able to afford bail.

In addition, according to Bureau of Justice Statistics, from 2000 to 2018, the jail population grew by 123,000 inmates; 95% of this increase (117,700 inmates) were people who had not been convicted of a crime (Josephson, 2018).

Often, even when a defendant can afford bail, some amount of time is required to accumulate the money. For example, 70% of the defendants who paid bail in New York City in 2017, had to spend some time in jail (Stringer, 2018).

Of course, there were other reasons for some of these defendants to be in jail; for example, some of these unconvicted detainees included (1) mentally ill people awaiting placement in mental facilities, or (2) defendants who were remanded (i.e., not given any bail). In addition, as noted previously, in some cases, a judge will set a high bail amount to keep the person in jail for public safety (Geng, 2018).

Obviously, the smaller the amount of time between arrest and trial of a defendant, the smaller the defendant's hardship in not being able to pay bail. However, even though the sixth amendment of the constitution guarantees the right to a **speedy trial** (again, a term that is not well defined), this will not often be the case. Different jurisdictions have different respective standards regarding the amount of time allowed between arraignment and trial. For example, in California, a felony defendant should be brought to trial within 60 days, while a misdemeanor defendant should be brought to trial within 30 days (if the defendant is in custody) or 45 days (if the defendant is not in custody) (Schwartzbach, n.d.).

The defendant could waive the right to a speedy trial in which case the standard time periods mentioned above do not apply. This waiving of a speedy trial may be desirable if extra time is required to prepare the defendant's defense.

In felony cases, there is usually a preliminary hearing following the arraignment in which the judge will decide if there is probable cause to go forward with the regular trial (Wallin, n.d.).

3.3.3 Trends in Bail Reform

There is a trend in recent years to relax the restrictions imposed by money bail. These relaxations are being accomplished using various state and federal laws. These laws can result in any of a variety of pretrial conditions for the accused. In many cases, these laws have resulted in making it easier for those who have been accused, but not convicted, to be released from jail. This is accomplished through either the setting of a lower amount for bail, or having the defendant released on his/her own recognizance (with possibly additional pre-trial restrictions).

Examples of some of the laws regarding pretrial conditions include: the Federal Bail Reform Act of 1966 (Bail Reform Act of 1966, n.d.), the Federal Bail Act of 1984 (The Bail Reform Act of 1984 (n.d.)), the Criminal Justice Reform Act of New Jersey, passed in 2014; the Money Bail Reform Act of California; and the New York State Law eliminating cash bail for many crimes, which went into effect from January of 2020.

Among other things, the Federal Bail Reform Act of 1966 resulted in the decreased use of money bail bonds and increased use of nonfinancial restrictions on defendants prior to trial.

The Federal Bail Reform Act of 1984 addressed the area of public safety by allowing a federal court to detain a defendant prior to trial if it could be demonstrated that the defendant posed a significant safety risk to the community. As such, this act moved in the opposite direction with respect to pretrial detention, as compared to the other acts mentioned above.

The 2014 Criminal Justice Reform Act of New Jersey, which became an amendment to the New Jersey Constitution and took effect on January 1, 2017, eliminated bail for many defendants (Balcerzak, 2020). With this law, a decision is made within 48 hours of arrest to either release or incarcerate the defendant and that decision is based upon (1) community safety, (2) an estimate of the probability that the defendant will appear for trial, and (3) an estimate of the probability that the defendant will obstruct justice (e.g., interact with witnesses in a way favorable to the defendant). (In most cases, the probabilities referred to in the second and third items are implicit in nature.) In effect, the financial status of the defendant is removed as a consideration with the elimination of money bail. Of course, the criteria for release stated above are subjective in nature.

The Money Bail Reform Act of California was signed in 2018 and took effect in 2019. This act eliminated the money bail system in California. Those accused of nonviolent misdemeanors could be released on their own recognizance within 12 hours of the booking process. For those accused of more serious crimes, a risk assessment associated with the release of the defendant until trial is made. The assessment is based on several factors including the alleged crime of the defendant; characteristics of the defendant such as prior criminal history, employment and housing situations, zip code of residence, and the safety concerns of the community. Based upon an algorithm which considers these factors, the defendant is placed into one of three categories of risk: low, medium, or high. Defendants placed in the low-risk category are to be released with the least restrictive, non-monetary conditions. Those placed in the medium-risk category could be either released or incarcerated until trial, depending on local standards. Finally, those in the high-risk category would remain in custody. The actual algorithm used to assign the defendant to one of the three risk categories is determined by each respective jurisdiction (Simon, 2018).

The new law addressing bail reform in New York State was passed in the spring of 2018. The law specified that defendants accused of misdemeanors or nonviolent felonies (except possibly for cases such as sex offenses or witness tampering) be released on their own recognizance (i.e., with no bail required). For other defendants, multiple forms of (noncash) bail were allowed, including unsecured and partially secured bonds. In addition, the amount of bail was set according to the defendant's ability to pay; however, the risk posed by the defendant to public safety was not considered. In fact, New York remains the only state that does not allow for the consideration of public safety in setting an amount for bail (Rahman, 2019).

Many researchers have criticized the risk assessment that is typically performed to consider public safety in the setting of a bail amount because of its capacity to support racial and other biases.

3.3.4 Advantages and Disadvantages of Keeping a Defendant Incarcerated Prior to Trial

Advantages of keeping a defendant in jail until trial include the following: (1) the defendant is assured of being present for the trial and (2) the defendant will not be committing (possibly) additional crimes. (Figueroa (2018) notes that problems with illegal drugs are a major reason for defendants out of jail on bail to commit new crimes.) Disadvantages associated with keeping a defendant in jail until trial include: (1) a possibly innocent person will be kept in jail; (2) the defendant will not be able to continue working a job, and may lose that job as a result; (3) the defendant will have a more difficult time getting their affairs in order given that a guilty verdict and a jail sentence occurs; and (4) incarceration of a defendant is expensive—for example, the cost to New York City of incarcerating those who cannot afford bail prior to their trial is approximately $100 million annually (Stringer, 2018).

There are other disadvantages, at least to the defendant, of incarceration prior to trial. For example, defendants in pretrial detention are three times as probable to be sentenced to prison as those who can make bail (Lowenkamp, VanNostrand and Holsinger, 2013).

Given that tens of thousands of defendants can be (and typically are) affected by new laws which address pretrial conditions, it should not be surprising that bad outcomes can occur in each of the two cases: (1) reforms associated with more strict pretrial release conditions and (2) reforms associated with less strict pretrial release conditions.

For example, in July of 2015, Sandra Bland was stopped for a minor traffic violation in Waller County, Texas. A confrontation ensued and she was arrested and charged with assault. Unable to pay $500 in bail, Ms. Bland was found hanging in her jail cell three days after her arrest. Her death was ruled as a suicide, although there has been some dispute about this ruling (Hennessy-Fiske, 2015).

At the other end of the spectrum is the situation of Gerod Woodberry of New York City. Mr. Woodberry allegedly robbed banks in New York City four times: December 30, 2019; January 3, 2020; January 6, 2020; and January 8, 2020, before he was finally arrested on January 8. He was released the next day, on January 9, 2020, without bail under New York State's new bail reform law. The main reason why he was released without bail is that his crimes were classified as robbery in the third degree and grand larceny but without the use of a gun; hence, his crimes were classified as nonviolent in nature. After being released on January 9, he allegedly robbed a fifth bank the next day, January 10, and was arrested again the same day. Mr. Woodberry was quoted as saying:

I can't believe they let me out. What were they thinking?

See Tracy (2020), Goldberg (2020), and Celona, Feuerherd, and Weissman (2020) for this and additional information on Mr. Woodberry's case.

3.3.5 Current Approaches and Suggestions for the Bail Process

Often a judge will use a bail schedule as a guide in setting the bail amount. Some examples of typical bail amounts are shown in Table 3.1. These amounts are derived from (Bail Amounts by Crime…, 2019).

There is much variation in the amounts of money required for bail for a specific crime, depending on the jurisdiction. For example, for lower-income states like Utah, the bail amount for illegal possession of a controlled substance could be set as low as $500.

It has been reported that 40% of adult Americans do not have enough savings to cover a $400 emergency (Backman, 2018). For

TABLE 3.1

Typical Amounts for Bail for Various Crimes

Misdemeanor Crimes	Typical Bail	Felony Crimes	Typical Bail
Illegal possession of a controlled substance	$2,500	Sexual Battery	$25,000
Public intoxication	$200 to $500	Accessory to murder in the first degree	$500,000
DUI on a suspended license	$2,500	Voluntary manslaughter	$100,000
Violation of a temporary restraining order	$15,000	Kidnapping	$100,000 (with much variation)
Battery against a peace officer	$2,500	First-degree robbery	$100,000

arrested Americans, one could expect this percentage to be much higher. One can therefore see why paying the money required for bail for a crime would be so difficult for many people.

In deciding about pretrial conditions for a defendant such as whether to keep said defendant incarcerated or not incarcerated, the amount required for bail and other restrictive conditions, a judge may consider any of several characteristics of the situation. Some of these characteristics have already been mentioned above in the discussion of the Criminal Justice Reform Act of New Jersey and the Money Bail Reform Act of California. These characteristics include the alleged crime of the defendant, community safety, the likelihood that the defendant will appear for trial, the likelihood that the defendant will obstruct justice, the likelihood that the defendant will commit another crime while out on bail, the likelihood of the defendant committing a technical violation related to his/her pretrial conditions, history of drug abuse, prior criminal history, employment and housing situations of the defendant, zip code of defendant's residence, nature of active community supervision at the time of arrest, the defendant's ability to pay, and the amount of congestion in the local jail (see Van Nostrand & Keebler, 2009; Bail Amounts by Crime…, 2019; Simon, 2018; and Rahman, 2019).

Obviously, some of these characteristics are more important than others in the determination of pretrial conditions for the defendant.

In addition, some of these conditions are dependent upon each other in a probabilistic sense; for example, prior criminal history and the likelihood that the defendant will commit another crime while out on bail are certainly dependent.

In most jurisdictions, judges will consider the situational characteristics listed above in a subjective manner through what might be termed as a subjective risk assessment. In a few jurisdictions, more formal approaches are used such as described above in the Money Bail Reform Act of California, or as suggested with objective risk assessments (Cooprider, 2009). Whether a less formal or a more formal approach is used, certain principles should be followed in determining pretrial release conditions, as noted by Hopkins, Bains, and Doyle (2018):

1. Releasing defendants on their own recognizance should be the norm;
2. Burdens and restrictions on defendants should not exceed the true interests of the government;
3. Restrictions placed on the defendant should be based on the alleged crime, the likelihood of flight, and the pre-arrest criminal activity of the defendant;
4. The defendant should not be charged for pretrial services; and
5. Constraints on the defendant should be evidence-based.

A summary of the situational characteristics considered by various acts/laws is shown in Table 3.2.

TABLE 3.2

Summary of Situational Characteristics Considered by Judges in Setting Pretrial Conditions

Source	Characteristics
2014 Criminal Justice Reform Act of New Jersey	community safety, probability that defendant will appear for trial given that he/she is released, probability that defendant will obstruct justice if released
Money Bail Reform Act of California	alleged crime of defendant, prior criminal history, employment and housing situations, zip code of residence, community safety
Bail Reform Law of New York State	defendant's ability to pay

See Summers and Willis (2010) for additional research on pretrial risk assessment.

3.3.6 A Multiple Objective, Decision Theoretic Model for Selection of Pretrial Conditions

In this section, we present a model and an illustrative example of that model for the selection of pretrial conditions for a defendant. The model is based on the optimization of expected utility, where the utility function contains three attributes; one of the attributes represents the interests of the "government" (or "the people") while the other two attributes represent the interests of the defendant. The fact that two of the attributes are mainly of concern to the defendant and only one is mainly of concern to the government is not an indication that the government's concerns are of less importance than the defendant's concerns. The weights associated with the utility function affect the overall value structure for the decision and can result in the government's concerns being of more importance than the defendant's concerns, even with one attribute versus two.

Uncertainty, and its inherent risk, associated with the behavior of the defendant is represented through a Monte Carlo simulation of the situation.

The model is to be used for a single defendant. However, it could be easily extended to consider a whole group of defendants, as would be done when considering a new law for the situation.

3.3.6.1 Stakeholders

Decisions regarding pretrial conditions for defendants should be based on probabilities, benefits, and costs that are associated with specific outcomes. In addition, the various trade-offs between multiple performance measures should be considered.

For example, in deciding whether or not to release a defendant prior to trial, and, if released, the amount of bail required and which restrictive conditions, if any, to place on the defendant, the judge will at least implicitly consider things like the (1) respective probabilities associated with the defendant committing additional crimes, intimidating witnesses, and not showing for trial; (2) the financial situation of the defendant; (3) the defendant's prior criminal history; and (4) the defendant's history of drug abuse among other things. With respect to the set of probabilities noted in (1), Bechtel, Holsinger, and Lowenkamp (2017) note that none of the conditions of release from their review considered the probability associated with the event of re-arrest of the defendant.

In the model and illustrative example developed in this section, we consider two categories of stakeholders: (1) the government/state and (2) the defendant. The category of the government/state includes any direct victim(s) of the crime and the taxpayers of the jurisdiction; typically, in criminal proceedings, this category is referred to as "the people". The defendant refers to the person accused of the crime as well as the group of people directly connected to the defendant, such as immediate family. One thing to keep in mind regarding the defendant is that he/she may very well be innocent of the crime; hence his/her interests (including the interests of direct family) should be considered in the decision regarding pretrial disposition.

Each of the two stakeholders are concerned with one or more performance measures (or attributes). We use a multiattribute utility function to represent the trade-offs that should be considered among the performance measures by a "super stakeholder" (i.e., the judge who considers both the government/state and the defendant). Remember that such a multiattribute utility function considers both the multiple, conflicting, performance measures as well as the uncertainty and resulting risk of the situation.

3.3.6.2 Alternatives and Process Description

In the situation under consideration a judge must choose one of several alternatives related to pretrial conditions for a defendant. These alternatives could be (1) remand, (2) release of the defendant with no restrictions, (3) release of the defendant subject to house arrest through electronic monitoring, (4) release of the defendant subject to other restrictions, etc. The release of the defendant may or may not be dependent on the payment of money for bail. If the defendant is given the chance for release from detention, he or she may or may not be able to take advantage of this opportunity, depending on financial status. In reality, the defendant may need to spend some time in jail (raising funds) prior to release. For the model presented here though, we will assume that the defendant is able to raise the required funds immediately or remains in jail until the results of the trial are determined. The basic situation described here is illustrated in Figure 3.1.

Note that in Figure 3.1, the squares represent decision nodes and the circles represent outcome nodes. Remember however that in the case of Figure 3.1, decisions by the defendant, as indicated for

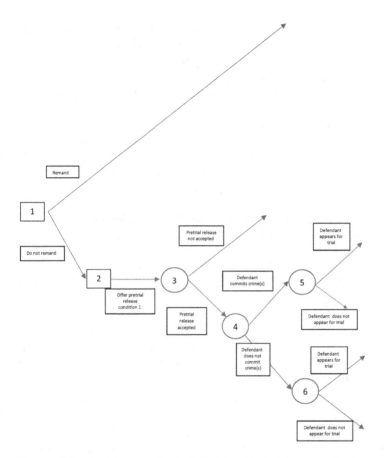

Figure 3.1 Decision tree for bail decision. A decision tree, flowing from left to right, as represented by nodes connected with directed arcs. There are two decision nodes (rectangles) and four outcome nodes (circles).

example by the arcs emanating from nodes 3, 4, 5, and 6 are represented as outcomes, in the typical terminology for decision trees, since these decisions are not under the control of the judge (decision maker).

As shown in Figure 3.1, the judge has an initial decision to make (as illustrated by the arrows emanating from node 1): remand (place under incarceration) or release. If the defendant is released, the judge will have a choice of any of several different options with respect to pretrial release; these options would be represented as arcs emanating from node 2. Due to space considerations, only one pretrial option is shown in Figure 3.1; however, other options would be shown as an equivalent set of arcs and nodes.

The arcs emanating from node 3, an outcome node, represent decisions of the defendant; namely, the defendant can either accept the pretrial alternative offered by the judge or reject it. Emanating from node 4 are arcs which correspond to the defendant either committing crimes when free prior to the trial or not committing any crimes. Finally, the arcs emanating from both nodes 5 and 6 indicate that the defendant will either appear for his/her trial or not appear.

The representation indicated by Figure 3.1 might be considered as simplified. For example, as has already been mentioned, after being offered pretrial release (node 2), the defendant may or may not have to wait some period prior to acceptance and release from jail. In this model, we are assuming that acceptance or rejection of the offer occurs immediately.

Another example of the simplified representation is that if the defendant commits a crime while released from jail prior to trial, he or she may or may not be apprehended for this new crime, which is not represented with this model. Instead the defendant will proceed to either appear or not appear for trial.

3.3.6.3 Input Parameters, Probabilities, and Control Variables for the Model

Associated with the decision tree of Figure 3.1 are several parameters. For example:

NPT = Number of alternatives related to pretrial release considered by the judge.

DAYSF = Number of days that the defendant will be free from arraignment to the end of a trial, given that he/she is released pretrial.

The number of arcs emanating from node 2 in an expanded version of Figure 3.1 would be equal to NPT, corresponding to the number of pretrial release conditions considered by the judge.

Some of these parameters are costs. These costs can be incurred by either the government or the defendant. In the context of the discussion here, the government represents everyone not associated with the defendant—that is, the taxpayer, victims of the defendant, etc. These costs are denoted by:

CINCG = Cost per day to the government to incarcerate the defendant,

CPT_i = Cost per day to the government for imposition of pretrial restrictive conditions under release alternative i, for i = 1, ..., NPT,

CNA = Cost to the government associated with nonappearance of the defendant for trial,

CAC = Cost to the government associated with additional crimes committed by defendant while on bail, given that additional crimes are committed,

$CDEFPT_i$ = Cost to the defendant associated with "inconvenience" due to the adherence to pretrial restriction i imposed by the government, for i = 1, ..., NPT,

CBAIL = Monetary cost to defendant of bail paid.

Some of the costs noted above would be relatively easy to determine. Others would be much more difficult. However, as noted previously, the judge would be considering these costs in an implicit fashion in this decision situation anyway, even if a less formal approach than the one described here were being used.

CINCG, cost per day to the government to incarcerate the defendant, and CPT_i would be relatively easy to estimate from existing data. For example, for CPT_i, if the pretrial restrictive condition involved regular reporting by the defendant, then probation officer salaries would be part of the consideration.

CNA, the cost to the government associated with nonappearance of the defendant, and CAC, the cost to the government for additional crimes committed by the defendant, would be much more difficult to determine. Each of these parameters would involve things like the cost associated with a search for the defendant. One should keep in mind that with respect to the model to be presented, each of these, as well as other parameters, could be represented as random variables.

The cost to the defendant associated with adherence to a pretrial restriction ($CDEFPT_i$) could be considered as whatever the defendant would pay (in dollars) to not adhere to the restriction; this would be in comparison to no restriction. For example, the restriction could be that the defendant travels to a location on a weekly basis to check in with a court official; the maximum amount that the defendant would pay to avoid this weekly trip would correspond to the value for $CDEFPT_i$. An alternative approach for computing this cost would be to consider the amount of time (say in hours) required by the defendant to fulfill the requirement, and then multiply this number of hours by the value of time (in dollars) of the defendant. If the pretrial restriction has to do with house arrest, using an electronic monitoring device, then a different approach might be required to estimate this cost.

With respect to the cost for bail/bond to the defendant (CBAIL), as noted previously the defendant will typically (1) pay the entire amount and then receive that amount back later assuming he/she shows for trial or(2) pay 10% of the amount to a bail bond company, but not receive any of the 10% back. In either case, in the model presented here we will assume that the cost to the defendant will be 10%, since even if he/she pays the full amount there will be inconvenience associated with not having the funds temporarily.

Finally, there are various probabilities that could be associated with the decision tree of Figure 3.1:

PNA_i = Probability of nonappearance of the defendant for trial, given that he/she is released under the alternative of pretrial release condition i,

PC_i = Probability that the defendant will commit additional crimes under the alternative of pretrial release condition i,

$PACC_i$ = Probability that the defendant will accept pretrial release condition i, if offered; if not accepted the defendant will remain incarcerated until trial.

Using information about the defendant, the judge can estimate various respective probabilities associated with (1) the defendant committing crime(s) prior to trial and (2) the defendant not showing for his/her trial. To keep the model relatively simple, these events of committing crimes and not showing for trial are assumed to be independent. In addition, these event probabilities can be based on the pretrial release conditions imposed. For example, very restrictive conditions on the defendant would result in relatively lower probabilities associated with the commission of crime(s) and of not showing for trial. These various probabilities could be either objective or subjective in nature. For example, if it is determined that for previous defendants with similar pretrial release conditions, prior criminal history, employment history, etc., 20% fail to show for trial, then one could estimate a probability of .2 for this defendant not showing for trial.

Note that the total number of mutually exclusive alternatives considered by the judge is NPT + 1.

Now, to formulate an optimization model for this situation, decision variables must be defined. In this case the decision variables correspond to whether the judge decides to remand the defendant, and, if released, the decision of choosing one of the pretrial release conditions. Thus, the decision variables would be:

x_R = 1, if remand is chosen; 0, otherwise.
x_{Pri} =1, if prerelease alternative i is chosen; 0, otherwise, for i = 1, 2, ..., NPT.

Since the judge can only take one of the actions regarding the defendant, we have the following constraint:

$$x_R + \sum_{i=1}^{NPT} x_{Pri} = 1$$

Each path shown in the decision tree of Figure 3.1 is mutually exclusive—i.e., only one can occur as a result of the judge's decision and the various outcomes that can occur. For example, the path corresponding to nodes 1-2-3-4-6, and then following the upper branch from 6 represents:

> The judge offering pretrial release condition 1 to the defendant, the defendant accepting this, the defendant not committing additional crimes until trial, and the defendant appearing for trial.

3.3.6.4 Attributes for the Multiattribute Utility Function

Now, there are several different performance measures that one might consider in making the decision corresponding to the situation described here. We want to make sure that we consider the various stakeholders (two in this case) and their objectives. One approach would involve employing the following set of attributes (or performance measures):

CGOV = Cost to the government associated with the decision made regarding bail,

CINB = Cost associated with inconvenience (as a result of an imposed pretrial restriction) + the cost associated with bail to the defendant,

DFREE = Number of days of non-incarceration for the defendant as a result of release of the defendant prior to trial.

CGOV is of course associated with the stakeholder that we identified as the government. This attribute consists of four types of costs: the cost for incarceration of the defendant, the cost for imposition of pretrial restrictions, the cost associated with nonappearance of the defendant at trial given that he/she does not appear, and the cost associated with crimes committed by the defendant while out on bail, prior to his/her trial. In addition, if the defendant does not show for trial, any bail that the defendant paid could be thought of as a negative cost (i.e., income to the government).

CINB and DFREE are attributes mainly of interest to the defendant. CINB is the sum of two costs, $CDEFPT_i$ and CBAIL, which were discussed above. In valuing DFREE, the judge, as the "super decision maker" must consider the comparison to incarceration and the direct and social costs associated with that incarceration.

For any specific decision made by the judge ($x_R = 1$, or $x_{Pri} = 1$, for one value of i = 1, 2, ..., NPT), the values associated with CGOV, CINB, and DFREE can be uncertain in nature as a result of the uncertain behavior of the defendant (as represented by the probabilities: PNA_i, PC_i, and PACCi for i = 1, 2, NPT) and the uncertain values of some of the other model parameters as a result of their representation as random variables.

3.3.6.5 Criterion, Alternatives, Parameters, and Multiattribute Utility Function for the Example

One criterion that could be employed in the evaluation of the judge's decision would be the maximization of the expected utility, where the utility is a function of the three attributes: CGOV, CINB, and DFREE:

Maximize EU (CGOV, CINB, DFREE)

In this illustrative example given below, we will employ a Monte Carlo simulation to estimate the expected utility of each alternative decision. Since, with the simulation, we will only be obtaining estimates of expected utility, we will employ one out of several available approaches for ranking the alternatives: the approach which involves "all pairwise comparisons" (pages 387–391, Evans, 2017). This approach is appropriate because of the relatively small number of alternatives under consideration.

Let's consider a situation involving an arraignment for an alleged nonviolent felony. The judge is considering remand of the defendant, along with three other alternatives (NPT = 3) with respect to pretrial conditions:

1. Release of the defendant on their own recognizance,
2. Release of the defendant with the restriction of weekly check-ins with the probation department, and
3. Release of the defendant on bail of $10,000, with no other conditions.

Now, a typical amount of time from the arraignment to the start of a trial is 30–45 days (Schwartzbach, n.d.), and a felony trial will sometimes require several months to complete (Chapman, n.d.). So, let us assume that the time from arraignment to disposition of the case through trial (DAYSF, as defined above) is uniformly distributed with a minimum value of 60 days and a maximum value of 120 days.

Estimates of costs associated with incarceration of prisoners vary widely. For example, a Bureau of Prisons estimate (2018) of the cost for keeping a prisoner in a residential re-entry facility in FY 2017 was $88.52 per day, while the same estimate for a federal prison was $99.42 per day. Estimates of the cost per day for keeping a prisoner in a local jail in 2015 vary according to $48 in Cherokee County, Georgia; $49 in Dallas County, Texas; and $83 in Douglas County, Nebraska (Henrichson, et al, 2015). In our example here, we will assume a cost per day of $90 to keep a prisoner incarcerated, i.e., CINCG = $90.

The costs for the three pretrial conditions specified are assumed to be $CPT_1 = 0$, $CPT_2 = \$390$, and $CPT_3 = 0$. Since the first and third pretrial conditions place no restrictions on the defendant and require no resources, these costs are just 0. The second pretrial condition requires a weekly check-in by the defendant; assuming 13 "check-ins", with each check-in requiring one hour of time for a court official who is paid $30 per hour, including overhead, we arrive at a cost of 13*$30 = $390.

If the defendant does not show for their court date, then typically the bail will be revoked, and a warrant will be issued for arrest. Bail-jumping is a crime in and of itself. The costs involved will include the administrative time for court officials as well as the time of police in searching for and re-arrest of the defendant. We will assume that this cost (denoted as CNA) is given by a random variable with a uniform distribution having a minimum value of $4,000 and a maximum value of $8,000.

There is also the possibility that the defendant will commit additional crimes if free on bail. We denote this cost associated with these additional crimes as CAC and assign its value as a random variable with a uniform distribution having a minimum value of $15,000 and a maximum value of $20,000.

The first step in developing a multiattribute utility function for this situation would be to determine the minimum and maximum values to allow for each attribute. These values would not have to be the exact minimum and maximum values, but they would need to just bound the actual minimum and maximum values.

Consider CGOV first. Without going into a lot of detail, the minimum value for CGOV would occur if the defendant is not incarcerated prior to trial, does not require any government resources prior to trial (i.e., is assigned pretrial release condition 1 or 3 above), does not commit any crimes while free on bail, and shows for his/her court date. This minimum value would be $0. The reader could verify this by assuming different alternative decisions and outcomes for the uncertain events and performing the relevant computations, relative to Figure 3.1.

The corresponding maximum value for CGOV would occur if the defendant were allowed to be free prior to trial by being released on his/her own recognizance, committed crimes at the highest level of cost according to the uniform distribution specified for this input ($20,000), and also did not show for their trial, also at the highest level of cost ($8,000). Hence, this maximum cost would be $28,000.

The minimum value for CINB would occur if the judge chose the first pretrial condition (to release the defendant on their own recognizance). This value would be $0.

The maximum value for CINB would be equal to $CDEFPT_2$ plus CBAIL, or $1000 + $1000 = $2000.

The minimum value for DFREE would be 0 days (this would be the worst outcome for the defendant for this attribute as DFREE represents the number of days of "non-incarceration"), while the maximum value for DFREE would be 120 days. This value of 120 days represents the maximum possible value for the number of days from arraignment to the end of the trial (i.e., the maximum possible value for U (60 days, 120 days)).

In order to keep this illustrative example relatively simple, we will assume that the multiattribute utility function is a scaled additive function (from 0 to 1), with the individual attribute utility functions being linear. Representing the attributes as X_1, X_2, and X_3 we have:

X_1 = CGOV,
X_2 = CINB, and
X_3 = DFREE.

We will assume a scaled, additive utility function:

$$u(x_1, x_2, x_3) = \sum_{i=1}^{3} k_i u_i(x_i)$$

,

where u_1, u_2, and u_3 are the individual attribute utility functions for x_1, x_2, and x_3, respectively, and k_1, k_2, and k_3 are the corresponding scaling constants.

Given the minimum and maximum values for each of the attributes and the fact that the individual attribute utility functions are linear, we can express these individual attribute utility functions as:

$u_1(x_1) = -.000035714x_1 + 1$,
$u_2(x_2) = -.0005x_2 + 1$, and
$u_3(x_3) = .0083333x_3$

At the maximum values of attributes X_1 and X_2, the corresponding respective utility values are each 0, and at their corresponding minimum values the respective utility values are each 1. At the maximum value for the third attribute, X_3, the corresponding utility value is 1; when X_3 has a value of 0, its minimum value, the corresponding utility value is 0. These utility values correspond to the fact that we want to minimize both X_1 and X_2 and maximize X_3.

We assume values for the scaling constants as follow:

$k_1 = .4$,
$k_2 = .3$, and
$k_3 = .3$.

Note that the form of the multiattribute utility function (additive, instead of say multilinear or multiplicative), the form and specifications of the individual attribute utility functions: u_1, u_2, and u_3 (linear in this example), and the values for the scaling constants (k_1, k_2, and k_3), would all be determined through an assessment process, as alluded to in Chapter 2 of this book, and in other books (for examples, see pages 236–263 of Evans (2017), and pages 261–271 of Keeney and Raiffa (1993)).

3.3.6.6 The Simulation Model and Results

The Monte Carlo simulation model employs a process-oriented simulation software package. Each replication of the model involves the generation of a single entity which flows through a network of modules. The branching from one module to the next is accomplished in either a deterministic fashion (depending on the value of a control variable which represents one of the four policies regarding bail listed above) or in a probabilistic way, modeling (1) whether the defendant accepts the pretrial release alternative offered by the judge, (2) whether the defendant commits additional crimes while free on bail, and (3) whether the defendant shows for their trial if released prior to trial. Random variates are generated for (1) the three probabilistic branching situations listed above (i.e., according to the probabilities given by $PACC_i$, PC_i, and PNA_i, for i = 1, ..., NPT); and (2) the uncertain quantities associated with DAYSF, CAN, and CAC.

The main output of the simulation is an estimate of the expected utility associated with one of the four policies modeled with the current simulation run. In addition, the 95% confidence interval associated with the expected utility is output by the model. There is a large body of literature involving the choice of a best alternative using a simulation (see pages 371 through 412 of Evans (2017) or pages 548 through 574 of Law (2007)); a reason for the difficulty here is the fact that we obtain only an **estimate** of the criterion value (i.e., value for expected utility) used for making a choice. This subject area is beyond the scope of this book; suffice it to say, however, that since there are only four policies in this example, and that 100 independent replications of the simulation are made for each policy, we can easily distinguish between the expected utility estimates for the four policies.

In addition to expected utility, the model also outputs an estimate for the expected values of each of the attributes: CGOV, CINB, and DFREE.

TABLE 3.3
Inputs to the Simulation Model

Input	Definition	Value (for this example)
NPT	Number of alternatives under consideration for pretrial release	3
DAYSF	Number of days that defendant will be released under a pretrial release alternative	U (60 days, 120 days)
CINCG	Cost per day to incarcerate the defendant to the government	$90
CPT_i, for i = 1, 2, 3	Cost to the government to implement pretrial release condition i	$0, $390, $0
CNA	Cost to the government associated with nonappearance of the defendant at trial	U ($4000, $8000)
CAC	Cost to the government associated with additional crimes committed by defendant while on bail, given that additional crimes are committed	U ($15000, $20000)
$CDEFPT_i$, for i = 1, 2, 3	Cost to the defendant associated with "inconvenience" due to the adherence to pretrial restriction i	$0, $1000, $0
$CBAIL_i$, for i = 1, 2, 3	Monetary cost to defendant of bail paid under pretrial release alternative i.	$0, $0, $1000
PNA_i, for i = 1, 2, 3	Probability of nonappearance of the defendant for trial, given that he/she is released under the alternative of pretrial release condition i	.2, .1, .05
PC_i, for i = 1, 2, 3	Probability that the defendant will commit additional crimes under the alternative of pretrial release condition i	.2, .1, .2
$PACC_i$ for i = 1, 2, 3	Probability that the defendant will accept pretrial release condition i, if offered; if not accepted the defendant will remain incarcerated until trial	1., .9, .6
x_R	Decision variable set to 1 if remand is chosen, 0 otherwise	Set according to policy chosen
X_{Pri}, for I = 1, 2, 3	Decision variable set to 1 if pretrial release alternative i is chosen, 0 otherwise	Set according to policy chosen

A summary of the inputs to the simulation model is shown in Table 3.3.

For access to the simulation model developed for this example, contact the author.

The four policies: remand (policy 1), release on own recognizance (policy 2), release on condition of weekly check-ins (policy 3), and release on bail money (policy 4) were simulated under two different sets of preferences (as represented by the three scaling constants of the utility function) and under two different sets of probabilities associated with the defendant committing additional crimes while free prior to the trial (i.e., the PC_i values) and with the defendant not appearing for trial (i.e., the PNA_i values). The combinations of scaling constants and parameter values yield four different sets of results, as shown in Table 3.4.

Table 3.4 illustrates that Policy 2 (release on own recognizance) is ranked first for two of the four parameter settings shown, while Policy 1 (remand) and Policy 3 (release on condition of weekly check-in) are ranked first on one of the parameter settings each. Note that the various rankings shown are a result of not only the parameters that were varied, but also of values for the other parameters, such as the cost associated with the committing of additional crimes (CAC) and the cost associated with non-appearance of the defendant at trial. For example, as the cost associated with committing additional crimes and the cost associated with non-appearance at trial increase, Policy

TABLE 3.4
Estimates of Expected Utilities and Associated Policy Rankings Under Two Sets of Scaling Constants and Two Sets of Probabilities. (Policies are listed from most to least preferred, with expected utilities shown in parentheses following the policies.)

Parameter Settings	$k_1 = .4, k_2 = .3, k_3 = .3$	$k_1 = .6, k_2 = .2, k_3 = .2$
$PC_1 = .6, PC_2 = .3,$	Policy 2 (.71)	Policy 3 (.74)
$PC_3 = .3$	Policy 3 (.69)	Policy 4 (.72)
$PNA_1 = .6, PNA_2 = .3,$	Policy 4 (.67)	Policy 2 and Policy 1 (.63)
$PNA_3 = .3$	Policy 1 (.58)	
$PC_1 = .8, PC_2 = .5,$	Policy 2 (.64)	Policy 1 (.63)
$PC_3 = .5$	Policy 3 and Policy 4 (.61)	Policy 4 (.62)
$PNA_1 = .8, PNA_2 = .5,$	Policy 1 (.59)	Policy 3 (.60)
$PNA_3 = .5$		Policy 2 (.52)

1 (remand) becomes more attractive. The overall ranking of all policies is a function of all the probabilities, costs, and other parameters input for the model.

3.3.6.7 Final Discussion of the Model

The purpose of this example is to illustrate how simulation and multiple objective methodology (as embodied by the use of a multiattribute utility function) can be used to determine a policy with respect to bail reform. The modeling approach allows for explicit consideration of all relevant costs and probabilities, as well as the trade-offs between the various objectives of interest to the two main stakeholders: the "government" (or the "people") and the "defendant" (including those associated with the defendant).

There are many variations on the model presented here. One group of variations would be concerned with making the model more accurate. Examples of these variations would be (1) a consideration of the dependence between the events of the defendant committing additional crimes while free on bail and the defendant not appearing for trial, (2) a delay between the decision of the judge with respect to a pretrial alternative and the decision of the defendant in whether to accept the judge's decision, and (3) the modeling of a possible plea agreement by the defendant prior to the start of trial. These additional features of the model would not be difficult to represent through appropriate modifications of the simulation model.

Another variation on the model would be the consideration of additional inputs to the model. An example of such an input would be the probability of defendant being guilty of the alleged crime, the conditional probabilities of the defendant being found guilty/innocent given that he/she is guilty/innocent. Although it may not be appropriate to do so, many judges would consider the probability of guilt at least implicitly in making their pretrial decisions.

At least some of the input data used in the simulation runs described in this section may very well be inaccurate or even unrealistic. Hence, prior to using a model such as this for decision-making, additional research should be done to firmly establish these data values, even if they are input as random variables. Of interest here would be the inputs associated with the costs and probabilities of additional crimes committed and non-appearance for trial (CAC, CNA, PC, and PNA) since these values would have a large effect on whether the defendant would be released prior to trial. Another input of primary importance would be the multiattribute utility function used to represent the trade-offs and risk preferences that the stakeholders would be willing to make or accept. Assessment procedures, involving questions of and answers from a decision maker, are available to determine an accurate multiattribute utility function.

Related to the accuracy of the input data for the model would be the concept of additional experimentation. The experimentation reported in this section of the book was extremely limited in nature. Given the relatively large number of difficult-to-estimate input parameters, the use of experimental design methodology may be appropriate (Chapter 12 of Law (2007)).

Finally, and this may very well be the most important point of this section, the model along with the variations described in this section could easily be extended to consider new policies or laws with respect to bail reform. Instead of generation of one entity as input to the process model for each replication, many entities, representing the new defendants over some period to be affected by the new policy, would be input for each replication. Further, groups of entities, representing groups of defendants with approximately the same characteristics could be input for each replication, and in fact an optimization could be performed which would address the assignment of each particular defendant group to a particular pretrial alternative.

USING MULTIPLE OBJECTIVE ANALYTICS TO IMPROVE FEDERAL LAWS

4.1 INTRODUCTION

The major crime bills of the last 60 years include the Omnibus Crime Control and Safe Street Act of 1968, the Anti-Drug Abuse Act of 1986 (HR 5484, 1986), the Violent Crime Control and Law Enforcement Act (VCCLEA) of 1994, and the Criminal Justice Reform Act of 2018. This chapter addresses the latter two acts. The VCCLEA is often referred to as the Clinton Crime Bill since President Clinton was an advocate of the bill. The Criminal Justice Reform Act is often referred to as the First Step Act.

Following this introduction, we discuss the VCCLEA in the second section of this chapter. This act, costing approximately \$31 billion, was and still is the largest crime bill in the history of the United States. We discuss the motivation for the law, its elements, and the results of its implementation. In addition, we discuss its detractors and supporters, and the law's effect on crime and incarceration rates. A major portion of the law is the Community Oriented Policing Services (COPS) program; we provide a cost–benefit analysis of this program based on data provided from a GAO report.

In the third section of this chapter, we discuss the Criminal Justice Reform Act of 2018 (the First Step Act). A major portion of this section is a description of the Prisoner Assessment Tool Targeting Estimated Risks and Needs (PATTERN). This tool allows for the classification of prisoners according to their tendency toward violent behavior and likelihood of recidivism. Finally, an illustrative example involving the investigation of two controls for the First Step Act is provided. The example involves the use of a Monte Carlo simulation, interfaced with a multiattribute utility function to allow for the consideration of trade-offs among the various performance measures for this decision situation.

4.2 THE VIOLENT CRIME CONTROL AND LAW ENFORCEMENT ACT OF 1994

4.2.1 Impetus for the Law

Starting in 1987, into the early 1990s, the murder rate in the United States was increasing by about 5% each year, peaking at 9.8 deaths per 100,000 people in 1991 (see Table 1.1). Much of the violence resulted from the Crack Cocaine Epidemic, and many of the victims were young African Americans, living in the inner cities of the country (Lussenhop, 2016). As noted by Blumstein (1995) many of the criminals associated with these crimes were also young African Americans recruited for the illicit drug market. These racial inequalities were part of the motivation for the passing of the Violent Crime Control and Law Enforcement Act (VCCLEA) of 1994 (more commonly known as the Clinton Crime Bill). This Act was at the time, and still is, the largest (in terms of funding) crime bill in US history, costing approximately $31 billion. It contained 33 different subject areas (Rosenfeld, 2019). The more specific provisions of the law are drawn from these subject areas or titles.

4.2.2 Elements of the Law

Some of the main subject areas of the act, as specified in Rosenfeld (2019), are public safety and policing, prisons, crime prevention, violence against women, drug courts, death penalty, mandatory life imprisonment for persons convicted of certain felonies, applicability of mandatory minimum sentences in certain cases, firearms, youth violence, crimes against children, state and local law enforcement, victims of crime, and sentencing.

Some of the specific provisions of the act include the following:

1. $7.6 billion in funding was provided for the establishment of the office of Community Oriented Policing Services (COPS), which allowed for the hiring of approximately 100,000 additional police in cities throughout the country.
2. $9.7 billion in funding was provided to states for the building of new prisons.

3. The three-strikes law, which mandated life imprisonment for persons convicted of a violent felony or a serious drug offense after having been convicted previously of a violent felony on two or more separate occasions, was enacted.
4. Funding was provided for a database consisting of 12 years of data (from 1990 to 2001) on crime.
5. Truth-in-sentencing laws were enacted.
6. Block grants were provided for prevention of crime through, for example, recreational programs focusing on at-risk youth.
7. The Violence Against Women Act, which increased penalties for sexual abuse and expanded assistance to victims of sexual assault among several other things, was enacted.
8. Grants were provided to states and cities for the establishment of drug courts.
9. The federal death penalty was expanded to cover 60 offenses, including the murder of a federal law enforcement officer, extensive drug trafficking, drive-by-shootings, and carjackings resulting in death.

Rosenfeld (2019) provides more detailed descriptions of these provisions, along with several other provisions emanating from the law.

4.2.3 Implementation Results

The actual implementation of the VCCLEA did not always match the law as written. For example, as noted above, $9.7 billion was set aside to states for the building of new prisons; however, only $3 billion was actually spent during the seven years of the law's appropriations (Gest, 2019). At least part of the difficulty was that the funding to the states for prisons was tied to the truth-in-sentencing provision, which required that prisoners serve at least 85% of their stated sentences. This part of the law was called the "Violent Offender Incarceration and Truth-in Sentencing", shortened to VOI/TIS. In addition, many states had already begun to enact stricter sentences for certain crimes.

With respect to the provision of adding 100,000 additional police, a GAO Report (Community Policing Grants, 2005) found that in 2000, which was the peak year of expenditures for this provision, 17,000 additional officers were hired. However, from 1994 to 2001, contributions through the COPS program allowed for the hiring of 88,000 additional police, not the 100,000 as stated in the bill (Community Policing Grants, 2005).

The notion of community policing is defined as *an approach to policing that involves the cooperation of law enforcement and the community in identifying and developing solutions to crime problems* (page 39, Community Policing Grants, 2005). Examples of activities associated with community policing are *identifying crime problems by looking at records of crime trends and analyzing repeat calls for service, working with other public agencies to solve disorder problems, locating offices or stations within neighborhoods, and collaborating with community residents by increasing officer contact with citizens and improving citizen feedback* (page 43, Community Policing Grants, 2005).

It is apparent from the quotes in the previous paragraph that "community policing" is a nebulous concept that is difficult to measure. It is also apparent that the directives associated with the community policing grants could have been, and probably should be, more specific in nature.

4.2.4 Detractors and Supporters

The 1994 Act has both its detractors and its supporters. As is usually the case, whether one is a detractor, or a supporter depends upon how one values the various outcome measures of the act. In addition, because of its complex nature, it is often difficult to determine the effect of the bill on some measures, such as reduction in crime.

Much of the African American community are critics of the law because a relatively large portion of those incarcerated as a result of the VCCLEA are African Americans. In rebuttal, some politicians point to the fact that 10 Black mayors from major American cities urged the Congressional Black Caucus (CBC) to support the bill, which they eventually did. However, it is noted that while supporting the basic idea of the bill, the CBC introduced an alternative bill which included an additional $2 billion funding for drug treatment programs and $3 billion for early intervention programs (Eisen, 2019).

There has been much debate, and several studies about whether, and if so by how much, crime has been reduced as a result of the VCCLEA. As would be expected, most of the focus is on the effectiveness of the COPS program. Portions of the debate deal with the motivation of criminals and the motivation for persons to report crime.

4.2.5 Motivation of Criminals

With respect to the motivation of criminals, the utilitarian theory basically says that most criminals are rational and will choose actions which maximize their expected utility; therefore they will be less likely to choose crime as opposed to choosing legal means to obtain benefits if they are more likely to be arrested because of an increased police presence (Marvell and Moody, 1996).

A related theory is referred to as incapacitation (Ehrlich, 1972). This theory says that more criminals will be arrested if there are more police. This will lead to more criminals being incarcerated and not homeless. With fewer criminals homeless, there will be less crime.

Marvell and Moody (1996) discuss several flaws in each of these theories, as well as reasons why adding more police does not often lead to a reduction in crime. One reason is that police strategies such as reliance on car patrols and responding to 911 calls are not good strategies for crime reduction. Another reason mentioned by Marvell and Moody is that the utilitarian theory is not necessarily valid since criminals may not act in a rational way, at least according to the researchers in the area. For example, with more police, criminals may switch to less risky crimes; but to maintain their income levels, they may need to commit more of these less risky crimes. Hence, as noted by Marvell and Moody, more police can actually lead to more crime, depending upon how crime is measured.

4.2.6 The Law's Effect on Crime

Various studies have come to different conclusions concerning the bill's effect on crime and the effect, in general, of providing more funding for crime prevention. Examples of publications in this area include Zhao et al (2002), Worrall and Kovandzic (2007), Sullivan and O'Keefe (2016), and Mello (2018), who determined that support to local governments for crime prevention can be effective, especially during times of economic downturn. Others have been pessimistic about the effects of almost any overt action on reducing crime (Miller, 1989).

The GAO report mentioned earlier (Community Policing Grants, 2005) stated that:

> between 1993 and 2000, COPS funds contributed to a 1.3 percent decline in the overall crime rate and a 2.5 percent decline in the violent crime rate from the 1993 levels.

The GAO report stated that these effects held true even when other crime factors, such as economic conditions, were controlled for.

The GAO report also stated that, between 1993 and 2000, the overall crime rate declined by 26% and the violent crime rate declined by 32%. Hence, COPS funding contributed to 5% (or approximately 1.3% divided by 26%) of the overall crime rate reduction, and to a little more than 7% (or approximately 2.5% divided by 32%) of the violent crime rate reduction.

4.2.7 A Cost–Benefit Analysis of the COPS Program

The estimates for decrease in crime from the GAO report provide information that could be used in a rough cost–benefit analysis of the COPS program, given that some additional assumptions are made. The GAO report quoted above mentions only the decrease in crime from 1993 to 2000, but does not mention the decreases in 1994, 1995, etc. Also, given that there was a 1.3% decline in the overall crime rate and a 2.5% decline in the violent crime rate, due to the COPS program, one could estimate a 1.13% decline in the property crime rate from the COPS program. (The overall crime rate is heavily weighted towards the property crime rate, as discussed in Chapter 1; hence the weightings, based on numbers of crimes, as shown in Table 1.1, give a 1.3% overall value for the decrease, based on the weightings given to 1.13% (property crime rate) and 2.5% (violent crime rate).)

Now, if one assumes that the declines in violent crime rates and property crime rates ramp up in a linear fashion from 1993 to 2000, then these yearly percentage decreases (from the overall decrease for that year) as a result of the 1994 bill would be .3571429% for 1994, 2*.3571429% for 1995, 3*.3571429% for 1996, etc., up to 7*.3571429% = 2.5% for the year 2000 from the base year of 1993 for violent crime. (Note that 2.5%/7 = .3571429%.) Similarly, the yearly percentage decreases in property crime (from the overall decrease for that year) as a result of the bill would be .1614286% for 1994, 2*.1614286% for 1995, 3*.1614286% for 1996, etc., up to 1.13% for the year 2000 from the base year of 1993 for property crime. (Note that 1.13%/7 = .1614286%.)

As an example, let us consider the year 1997. Table 1.1 indicates that the number of murders decreased from 9.5 per 100,000 population in 1993 to 6.8 per 100,000 population in 1997. Or, considering the populations of those years, the number of murders decreased from 24,501 in 1993 to 18,199 in 1997—a decrease of 6,302 murders.

According to our reasoning, 4* .3571429% = 1.428% of this decrease (or about 90 murders) was a result of the 1994 bill.

Carrying this type of calculation over all years and over all categories of crimes, one could calculate that from 1993 to 2000, because of the 1994 crime bill: 718 murders, 1,305 rapes, 16,280 aggravated assaults, 19,842 robberies, 25,022 burglaries, 21,027 larcenies, and 13,642 vehicle thefts were averted. Multiplying these numbers by the costs of $7.3 million per murder, $185,558 per rape, $74,301 per aggravated assault, $57,300 per robbery, $11,154 per burglary, $1,821 per larceny, and $7,732 per vehicle theft (as discussed in Chapter 1), respectively, we arrive at a figure of $8.2 billion in costs averted through the crime bill as a result of the COPS program. This compares to a spending of about $5 billion for the COPS program (as opposed to the $7.6 billion budgeted) (Community Policing Grants…, 2005). Even if the assumptions made in this rough analysis are slightly off, the COPS program paid off if the GAO report was an accurate estimate of the percentage decrease in crime. Of course, a more complete analysis would involve the consideration of things such as costs associated with the imprisonment of additional criminals.

4.2.8 The Law and Incarceration Rates

One of the criticisms of the bill is that it led to a large increase in the incarceration rate in the country. This has supposedly occurred through the increase in funding to the states for building prisons, the increase in the number of federal prisoners, and the changes in the sentencing laws among other things. All other things being equal, a higher incarceration rate is certainly a bad thing. However, if it corresponds to having more violent offenders off the streets (and thereby reducing the violent crime rate), it is a good thing. Hence, the value that one places on the incarceration rate is a complex issue.

These increasing incarceration rates associated with the bill led to criticisms of some politicians, such as Bill Clinton, Joe Biden, and Bernie Sanders (see Rosenfeld (2019)) for their parts in getting the bill passed. This criticism of the bill focused in particular on the incarceration rate of young African Americans. Others, however, have disagreed with this assessment by noting that the trend in incarceration rate during the late 1990s was just a continuation of the trend from the late 1980s and early 1990s (Stanglin (2020) and Eisen (2019)). Others have noted that many states had already enacted laws that resulted in increased sentences for crimes prior to the enactment of the VCCLEA, and that these actions by the states caused whatever increase occurred in the incarceration rates (Gest (2019) and Sabol and Johnson (2019)).

President Clinton noted, in response to the previous criticism, that *90 percent of the people in prison too long are in state prisons and local jails* (not federal prisons). He also noted that, in praise of the bill: *because of that bill we had a 25-year low in crime, a 33-year low in the murder rate* (Farley, 2016).

As can be noted from the discussion above, the VCCLEA has a mixed legacy. Those who criticize the bill tend to overstate the bad effects and understate the good effects, and those who praise the bill tend to overstate the good effects and understate the bad effects. In addition, disagreements can occur as a result of disputes about the causes of crime.

4.3 CRIMINAL JUSTICE REFORM ACT OF 2018 (THE FIRST STEP ACT)

4.3.1 Impetus for the Law

Catherine Toney was the first woman released from prison under the First Step Act (more formally known as the Criminal Justice Reform Act of 2018). President Trump signed this act into law on December 20, 2018, and Ms. Toney was released in early 2019, after having served 16 years of a 20-year sentence for a drug offense. She had attempted and failed to receive an early release from prison multiple times prior to the enactment of the First Step Act. She was freed under the act's retroactive crack cocaine sentence reduction provision (Creed (2019) and Gornoski (2019)).

Tanesha Bannister was also released early because of the First Step Act. She was originally sentenced to life in prison for drug trafficking but was released in 2019 after serving 16 years in prison (Monk, 2019).

Various researchers have suggested that a conscientious approach to early release can be greatly beneficial. For example, in a study involving a large sample of federal offenders, Rhodes et al (2018) stated that prisoners' average length of stay could be reduced by 7.5 months with little impact on recidivism. Wakefield (2018) provides a critique to the Rhodes et al study.

The First Step Act addresses an unease over the last few decades that the nation's war on drugs had led to incarceration of too many nonviolent offenders; also, when they were released, many of these offenders were not adequately prepared to reenter society. In addition to the retroactive extension of the Fair Sentencing Act of 2010 (Fair Sentencing Act of 2010, 2010), the First Step Act ends the "three-strikes" provision (except for those convicted of a "prior serious felony") of the Clinton Crime Bill which resulted in a mandatory life sentence for those convicted of three drug crimes (Ho, 2018).

Blumstein (2011) noted that a primary reason for the growth in the prison population from the late 1970s to 2010 was a tenfold increase in the incarceration of drug criminals. Hence, addressing the issue of drug crimes was of great importance in reducing the prison population.

The retroactive extension of the Fair Sentencing Act of 2010, which addresses the inequities of sentencing for possession of powder cocaine vs. crack cocaine among other things, resulted in 2,387 Federal inmates receiving reductions in sentences averaging 71 months, or 26% of their sentences (The First Step Act of 2018, 2019, and Department of Justice, 2019a). The projection is that the First Step Act will result in a reduction of the federal prison population by about 53,000 prisoners over the next 10 years (Drusch, 2018b).

4.3.2 Provisions of the Law

For 2021, the First Step Act provides a budget of $409 million to the Department of Justice's Bureau of Prisons, an increase of $319 million over the 2020 budget. This increase of $319 million is to be used to (1) fund recidivism-reducing programs for inmates who are given access to pre-release custody in the community ($244 million), (2) expand drug treatment programs for inmates ($37 million), and (3) increase the availability of such recidivism reduction efforts like classes in vocational training and life skills and mental health treatment ($23 million). In addition, the act provides $90 million in 2020 to support its implementation (The First Step Act of 2018, 2019).

4.3.3 Criticisms of the Law

Two of the criticisms of the First Step Act are that (1) it affects only federal prisoners and (2) it focuses on "back-end reforms"—i.e., on federal prisoners who have already served the bulk of their sentences. About 87% of prisoners are held in state, as opposed to federal, prisons (Haynes, 2018). Of course, these criticisms might also be called "missed opportunities", to be corrected in the future. In addition, given the annual costs associated with crime in the United States, one might also argue that the provisions of the act could have been funded at a much higher level, resulting in even larger benefits. In comparison, funding provided for the Clinton Crime Bill was at a much higher level than that for the First Step Act.

Even though the First Step Act affects only federal prisoners, it is based on similar reforms done on the state level in states such as Kentucky, Georgia, South Carolina, and especially Texas. For example, in Texas, reforms similar to those instituted in the First Step Act were implemented and resulted in savings of more than $4 billion between 2006 and 2016, after an initial investment of $241 million in rehabilitation programs (Drusch, 2018a).

4.3.4 The Prisoner Assessment Tool Targeting Estimated Risks and Needs (PATTERN)

Another important aspect of the First Step Act (FSA) is a new approach for classifying prisoners according to forecasts of (1) their tendencies toward violent behavior and (2) their respective likelihoods of recidivism. (Caulkins et al (1996) among others have also developed approaches for predicting recidivism.) This approach, termed the FSA Risk and Needs Assessment System, has been named using an acronym, PATTERN (Prisoner Assessment Tool Targeting Estimated Risk and Needs). The forerunners of PATTERN, which were used by its developers as a base from which to start the development of PATTERN were the Bureau of Prison's BRAVO (Bureau Risk Assessment Verification and Observation) system and BRAVO-R which added the methodological feature of dynamics and the application feature of recidivism to the BRAVO tool.

PATTERN works by initially assigning incoming prisoners to specific categories of risk with respect to both general and violent recidivism and violent/serious misbehavior at the point of prison intake. The risk levels are defined as being minimum, low, medium, and high. Periodically, based upon the prisoner's current level of risk as well as other dynamic and static factors, the prisoner is assigned to specific amounts and types of programming (classes for rehabilitation) and provided with incentives and rewards. In addition, on a periodic basis, the determination of when a prisoner is to be transferred into prerelease custody or supervised release is made. The process just repeats over time for each prisoner (page 5, Department of Justice, 2019b). As mentioned, the initial assessment of risk categories and of programming takes place upon prisoner intake, while subsequent assignments occur seven months after intake and then yearly after that.

In one sense, PATTERN can be viewed as representing a stochastic process in which the prisoner moves from one state (representing the risks of recidivism and violent/serious misbehavior) to another, on a periodic basis. The movement from one state to another is a function of both static and dynamic factors (discussed in greater detail below), as well as the programming/rehabilitation classes that the prisoner willingly undergoes. The states associated with the process can be defined not only by risk categories, but also by location—e.g., in prison, in prelease custody, or supervised release. If one wants to represent the action of recidivism, PATTERN could be used to represent the transition from the supervised release back to prison.

As noted in the document describing PATTERN, eligible prisoners earn 10 days of credit for every 30 days of successful engagement in programming and other effective activities (page 6, Department of Justice, 2019b).

PATTERN was developed by Drs. Zachary Hamilton and Grant Duwe; they used a data set consisting of information about 278,940 inmates released from the Bureau of Prison facilities between 2009 and 2015. Static factors (or predictors) used by PATTERN include things such as gender, age at the time of first conviction, age at the time of assessment, an indicator of whether or not the crime of conviction was violent in nature, and an indicator of whether or not the prisoner was a sex offender. Dynamic factors used by PATTERN include the following: the inmate's participation and performance in the programs to which he or she was assigned, number of infraction convictions during the current incarceration, number of serious infraction convictions during the current incarceration, number of programs (educational, vocational, drug treatment, etc.) completed as measured according to an ordinal category, an indicator (yes or no) as to whether the inmate participated in federal industry employment during current incarceration, drug treatment programs completed according to an ordinal scale, the inmate's willingness to use income earned during incarceration to reimburse victim(s) of their crimes (page 45, Department of Justice, 2019b).

An important project that could be accomplished as a follow-on to the development of the PATTERN system would be the development of a simulation model. This model would predict, over time, for a federal prison, the numbers of prisoners contained within the various categories of risk; the numbers of prisoners granted early release, pre-release custody, and supervised release; and the number of prisoners arrested or returned to federal prison within three years of release (a measure of recidivism).

Viewed as a process-oriented simulation, prisoners would be represented as entities being routed from one state to another. Input to the simulation would include probabilities associated with the various levels of risk for violent misbehavior, general recidivism, and violent recidivism. For example, a minimum risk of violent recidivism might correspond to a probability of .1, while a high risk of violent recidivism might correspond to a probability of .7. Attributes (or descriptors) for the entities (prisoners) would be the values associated with the static factors and dynamic factors for each prisoner.

Output from the simulation would be the numbers of prisoners associated with various categories of risk (for violent misbehavior, and general/violent recidivism), and the number of released prisoners who commit a crime (thereby providing statistics on recidivism) over simulated time, say a ten-year period. The output would be probabilistic in nature, thereby requiring appropriate experimentation with the model to reach valid conclusions. This type of output, especially the output which gives the numbers of prisoners contained in the various risk-level categories, would be useful to the prison for planning purposes. One of the directives associated with the First Step Act is that prisoners in similar risk-level categories be housed together, so the output providing information on the numbers of prisoners at the various risk levels over time would be useful in capacity planning for the prison.

4.3.5 An Illustrative Example: Optimizing Over Controls for the First Step Act

Consider a situation in which we have some flexibility in two control variables associated with the First Step Act. Let us formulate an optimization problem involving uncertainty and its inherent risk associated with this situation. We will assume that a simulation model, like the hypothetical model discussed above, is available for use.

We denote our control variables for this optimization problem as:

> r = annual funding for drug treatment programs, classes in vocational training and life skills, and mental health treatment for those incarcerated in federal prisons, in millions of dollars, and

d = the number of days of credit for every 30 days of successful engagement in programming and other effective activities by the inmate.

As noted above, the values associated with these control variables for the current version of the First Step Act are $r = 60$ (millions of dollars) and $d = 10$ (days).

Our output variables (also called attributes) of interest for this optimization problem are:

x_1 = annual cost for drug treatment programs, classes in vocational training and life skills, and mental health treatment for those incarcerated in federal prisons, in millions of dollars,

x_2 = number of prisoners (in hundreds) released on an annual basis because of receiving d days of credit for every 30 days of successful engagement in programming and other effective activities, and

x_3 = number of prisoners (in hundreds) who recidivate on an annual basis of those prisoners who are released early because of their successful engagement in programming and other effective activities.

As noted from Chapter 1, there are several different definitions that we could use for recidivism. In addition, as noted above with respect to the discussion of PATTERN, we have the categories of general recidivism and violent recidivism. To keep our discussion here parsimonious, we will define x_3 as the number of prisoners (in hundreds) who are re-arrested (for either a technical violation or a new crime).

Now, clearly r, x_1, x_2, and x_3 would not stay constant from one year to the next. However, we will assume that we are looking at mean values and a steady-state analysis of the output from the simulation. Also, in a more general sense, we are considering a general policy analysis in this exercise; that is, we are considering an annual spending amount and an annual value for the number of days of credit for every 30 days of enrollment over several years.

It is also clear that with respect to preferences, any rational decision maker would want to minimize x_1, maximize x_2, and minimize x_3, individually; all other things being equal.

To keep the situation simple, we will assume that r is equal to x_1— that is, the funding for programs/classes is equal to the cost. These two values should be close to equal anyway.

In terms of functional relationships, one could assume that as r increases (keeping d at the same value), x_1 would of course increase; in addition, x_2 would possibly increase since the capacity of vocational training and life skills classes would increase, thereby allowing more prisoners to complete these programs/classes, at least up to the point of the demand for these programs/classes. x_3 would also possibly increase as a result of an increase in the number of prisoners released; of course, the increased spending on programs/classes could result in countervailing effect through an increase in the quality of the programs/classes offered to prisoners. We would, of course, rely on either the simulation model or expert opinion to provide the accurate outputs in this case; such outputs may very well be probabilistic in nature, as discussed below.

As d increases, keeping r at a constant value, one should expect that both x_2 and x_3 would increase in value. Again, we would use our hypothetical simulation model or expert opinion to estimate values for x_2 and x_3.

The criterion that we will use for this decision problem is the maximization of expected utility, where our multiattribute utility function will have three attributes: x_1, x_2, and x_3, as defined above. For any policy (defined by the values assigned for r and d), the value for x_1 will be deterministic in nature, but the values for x_2 and x_3 will be probabilistic, as determined through the application of our hypothetical simulation model. Another output from our simulation will be a utility function value (for one replication of the model), and an estimate of expected utility (for multiple replications of our model).

Suppose that we would like to investigate the following policies:

Policy 1:   r = 40, d = 9;
Policy 2:   r = 40, d = 10;
Policy 3:   r = 60, d = 9;
Policy 4:   r = 60, d = 10;
Policy 5:   r = 80, d = 9; and
Policy 6:   r = 80, d = 10.

Note that policies 1 and 2 allow for the lowest amount of funding ($40 million) for programs and classes, while policies 5 and 6 allow for the largest amount of funding. Also, policies 1, 3, and 5 are the most restrictive with respect to days of credit for every 30 days of programs/classes completed (9 days), while policies 2, 4, and 6 are the most generous with respect to this control (10 days).

The results associated with these policies, as mentioned earlier, will be obtained either through experimentation with our simulation model or through expert opinion. We will assume that this output will be given in terms of triangular distributions, and that in terms of recidivism, a rate/probability of recidivism for a randomly chosen released prisoner will be provided. The other output, number of prisoners released early, will just be represented as a number. The distributions associated with each policy are given below:

Policy 1: recidivism rate: TRIA (.4, .42, .45), early release prisoners no.: TRIA (3500, 4000, 4700),

Policy 2: recidivism rate: TRIA (.4, .42, .45), early release prisoners no.: TRIA (3900, 4300, 4900),

Policy 3: recidivism rate: TRIA (.34, .37, .41), early release prisoners no.: TRIA (4000, 4500, 5200),

Policy 4: recidivism rate: TRIA (.34, .37, .41), early release prisoners no.: TRIA (4300, 4700, 5400),

Policy 5: recidivism rate: TRIA (.32, .35, .38), early release prisoners no.: TRIA (4400, 4800, 5400),

Policy 6: recidivism rate: TRIA (.32, .35, .38), early release prisoners no.: TRIA (4700, 5200, 5600).

The triangular distribution is especially useful when relatively simple input from an expert is needed. That is, if input is needed to specify the parameters of a, b, and c for a triangular distribution, TRIA (a, b, c), the expert could be asked to provide values for the smallest possible value for the relevant quantity (a), the most likely value (b), and the largest possible value (c).

To keep our exposition straightforward, we will assume an additive, scaled multiattribute utility function in which the individual attribute utility functions: u_1, u_2, and u_3 are linear. Operationally, in terms of interfacing the multiattribute utility function with the Monte Carlo simulation, there would be no difficulty in using a more complex function.

Our multiattribute utility function in this case is given by:

$u(x_1, x_2, x_3) = w_1u_1(x_1) + w_2u_2(x_2) + w_3u_3(x_3)$, where the scaling constants are given by
$w_1 = .1$, $w_2 = .6$, and $w_3 = .3$.

The individual attribute utility functions are given by:

$u_1(x_1) = -.016667x_1 + 1.4999,$
$u_2(x_2) = .033333x_2 - .99999,$ and
$u_3(x_3) = -.1x_3 + 2.3.$

Note that each of these individual attribute utility functions are linear, with a worst value for each attribute giving a utility function value of 0 and a best value for each attribute giving a utility function value of 1. That is, approximately, we have $u_1(90) = 0$, $u_1(30) = 1$, $u_2(30) = 0$, $u_2(60) = 1$, $u_3(23) = 0$, and $u_3(13) = 1$.

A Monte Carlo simulation model is used to give an estimate of expected utility for each policy.

The number of replications of the model for each policy has been set to respective values that assure an adequate differentiation between the policies in terms of their expected utility values (see pages 385 through 412 of Evans (2017)).

The simulation is relatively straightforward and executes with the following sequence. For each independent replication of the model:

1. Generate a random variate representing the number of prisoners to be released under the policy being simulated.
2. Generate a random variate representing the probability (according to the policy being simulated) that an early-release prisoner will be a recidivist.
3. For each of the early-release prisoners determine whether that prisoner is a recidivist according to the probability determined from the previous step.
4. If the prisoner is a recidivist, increase the number of recidivists by 1.
5. After modeling each early-release prisoner as a recidivist or not, compute values for the individual attribute utility function and for the multiattribute utility function.

Note that for each independent replication of the model for a specific policy, a different number of early-release prisoners can be computed, according to the distribution of early-release prisoners for this policy. In addition, for each independent replication, a different probability for an early-release prisoner being a recidivist is computed, and a different value for utility is computed.

After all replications for a policy have been executed, estimates of expected utility, expected values for cost, number of early-release prisoners, and number of recidivists are computed, as well as 95% confidence intervals for these quantities.

The results associated with the simulation of each of the six policies are given below. These results show, in order for each policy, its ranking among the policies, expected utility, expected cost (millions of dollars on an annual basis), expected number of early-release prisoners on an annual basis, and the expected number of recidivists out of the early release prisoners on an annual basis.

- Policy 1, 6th, .473, 40, 4084, 1725.
- Policy 2, 5th, .494, 40, 4381, 1851.
- Policy 3, 4th, .544, 60, 4584, 1708.
- Policy 4, 2nd, .565, 60, 4816, 1795.
- Policy 5, 3rd, .555, 90, 4881, 1705.
- Policy 6, 1st, .583, 90, 5181, 1809.

For example, Policy 6 was the first ranked policy with an expected utility of .583, an expected cost of $90 million, an expected number of early-release prisoners on an annual basis of 5181, and an expected number of recidivists on an annual basis out of those released early of 1809.

The reader should note that the rankings and expected utilities are a function of the multiattribute utility function. As mentioned in Chapter 2, this function would normally be determined through the use of a detailed assessment process in which the decision maker would answer questions concerning their trade-offs over hypothetical multidimensional, probabilistic outcomes. Different decision makers would accept different trade-offs, thereby leading to different respective utility functions and possibly different rankings for the policies.

The simulation model can be obtained from the author upon request.

REFERENCES

A Framework for Evidence-Based Decision Making in State and Local Criminal Justice Systems. (2017, June). *An initiative of the national institute of corrections*. 4th ed: *A continued work in process*. Retrieved on June 30, 2020 from https://s3.amazonaws.com/static. nicic.gov/Library/033067.pdf.

Abdul-Quader, A.S., Feelemyer, J., Modi, S., Stein, E.S., Briceno, A., Semaan, S., Horvath, T., Kennedy, G.E., & Des Jarlais, D.C. (2013). Effectiveness of structural-level needle/syringe programs to reduce HCV and HIV infection among people who inject drugs: A systematic review. *AIDS and Behavior, 17*(9), 2878–2892.

Access to Clean Syringes. (n.d.). Centers for Disease Control and Prevention. Retrieved on June 21, 2020 from https://www.cdc.gov/policy/hst/hi5/cleansyringes/.

Ackerman, M. (2019, December 15). The pros and cons of needle exchange programs. *Recovery.org*. Retrieved on June 23, 2020 from https://www.recovery.org/the-pros-and-cons-of-needle-exchange-programs/.

Ackoff, R.L. (1979). The future of operational research is past. *Journal of the Operational Research Society, 30*, 652–658.

All Data Collections. (n.d.). Bureau of Justice Statistics. Retrieved on April 28, 2020 from https://www.bjs.gov/index.cfm?ty=dca.

Alper, M., Durose, M.R., & Markman, J. (2018, May 23). 2018 update on prisoner recidivism: A 9-year follow-up period (2005–2014). Bureau of Justice Statistics. Retrieved on April 15, 2020 from https://www.bjs.gov/index.cfm?ty=pbdetail&iid=6266.

Aos, S., & Drake, E. (2010). *WSIPP's benefit–cost tool for states: Examining policy options in sentencing and corrections*. Olympia: Washington State Institute for Public Policy.

Auerhahn, K. (2002). Selective incapacitation, three strikes, and the problem of aging prison population: Using simulation modeling to see the future. *Criminology and Public Policy, 1*(3), 353–388.

Auerhahn, K. (2004). California's incarcerated drug offender population, yesterday, today, and tomorrow: Evaluating the war on drugs and proposition 36. *Journal of Drug Issues, 34*, 95–120.

Auerhahn, K. (2008). Using simulation modeling to evaluate sentencing reform in California: Choosing the future. *Journal of Experimental Criminology*, *4*, 241–266.

Backman, M. (2018, May 23). Guess how many Americans couldn't cover a $400 emergency. *The Motley Fool*. Retrieved on March 30, 2020 from https://www.fool.com/retirement/2018/05/23/guess-how-many-americans-couldnt-cover-a-400-emerg.aspx.

Bae, K.-H.G., & Evans, G.W. (2019). An overview of analytics for the design and operation of criminal justice systems. In G.W. Evans, W.E. Biles, & K.-H.G. Bae (Eds.), *Analytics, operations, and strategic decision making in the public sector* (pp. 208–225). Hershey, PA: IGI Global.

Bail amounts by crime—How much is bail? —Average bail prices. Bail Bonds Network. (2019, September 25). Retrieved on January 20, 2020 from https://bailbondsnetwork.com/bail-amounts-how-much.html#1.

Bail Reform Act of 1966. (n.d.). University of Pretrial. Retrieved on August 9, 2020 from https://university.pretrial.org/testing/glossary/entry?GlossaryKey=b8ab5066-68d5-468e-a861-f7d0242e2e84#:~:text=Final.%20The%20first%20major%20reform%20of%20the%20federal,could%20impose%20only%20if%20non%20financial%20release%20.

Balcerzak, A. (2020). New Jersey didn't see crime spike after 2014 reform. Will New York be the same? *NorthJersey.com*. Retrieved on August 28, 2020 from https://www.northjersey.com/story/news/2020/01/08/nj-bail-reform-didnt-see-spike-crime-ny-same/4396763002/.

Barnett, A. (1988). Misapplication reviews: Crime news. *Interfaces* *18*(3), 110–115.

Barnett, A., Caulkins, J.P., & Maltz, M.D. (2000). Crime and justice. In Saul I. Gass & Carl M. Harris (Eds.), *Encyclopedia of operations research and management science*, 2nd ed. Boston, MA: Kluwer Academic Publishers.

Basadur, M.S., Ellspermann, S.J., & Evans, G.W. (1994). A new methodology for formulating ill-structured problems. *Omega*, *22*(6), 627–645.

Bechtel, K., Holsinger, A.M., Lowenkamp, C.T., & Warren, M.J. (2017). A meta-analytic review of pretrial research: Risk assessment, bond type, and interventions. *American Journal of Criminal Justice*, *42*, 443–467.

Berenji, B., Chou, T., & D'Orsogna, M.R. (2014). Recidivism and rehabilitation of criminal offenders: A carrot and stick evolutionary game. *PLOS One*, *9*(1), e85531. Retrieved on July 24, 2020 from https://journals.plos.org/plosone/article?id=10.1371/journal.pone.0085531.

Berk, R. (2008). How you can tell if the simulations in computational criminology are any good. *Journal of Experimental Criminology*, *4*, 289–308.

Best Definition of Analytics (n.d.). INFORMS web site. Retrieved on August 21, 2020 from https://www.informs.org/About-INFORMS/News-Room/O.R.-and-Analytics-in-the-News/Best-definition-of-analytics.

Black's Law Dictionary, 11th ed. (2019). Bryan A. Garner, Editor. Toronto: Thomson-Reuters.

Blumstein, A. (1995). Youth violence, guns, and the illicit-drug industry. *Journal of Criminal Law and Criminology*, *86*(1), 10–36.

Blumstein, A. (2002). Crime modeling. *Operations Research*, *50*(1): 16–24.

Blumstein, A. (2007). An OR missionary's visits to the criminal justice system. *Operations Research*, *55*, 14–23.

Blumstein, A. (2011). Bringing down the U.S. prison population. *The Prison Journal*, *91*(3), 12S–26S.

Blumstein, A., & Cohen, J. (1987). Characterizing criminal careers. *Science*, *237*(4818), 985–991.

Blumstein, J.F., Cohen, M.A., & Seth, S. (2007). *Do government agencies respond to market pressures? Evidence from private prisons*. Nashville, TN: Vanderbilt University Press.

Blumstein, A., & Larson, R. (1969). Models of a total criminal justice system. *Operations Research*, *17*, 199–232.

Bohigian, H.E. (1977). Simulation modeling of the criminal justice system and process. *Proceeding of the 1977 Winter Simulation Conference*, pages 247–256. December 5–7, 1977, Gaithersburg, MD

Bond, S., Carlson, K., & Keeney, R.L. (2010). Improving the generation of decision objectives. *Decision Analysis*, *7*(3), 238–255.

Bryant, S. (2020, February 22). The business model of private prisons. *Investopedia*. Retrieved on April 3, 2020 from https://www.investopedia.com/articles/investing/062215/business-model-private-prisons.asp.

Buede, D.M. (1986). Structuring value attributes. *Interfaces*, *16*, 52–62.

Burdeen, C.F. (2016, April 12). The dangerous domino effect of not making bail. *The Atlantic*. Retrieved on May 25, 2020 from https://www.theatlantic.com/politics/archive/2016/04/the-dangerous-domino-effect-of-not-making-bail/477906/.

Bureau of Justice Statistics: Courts. (n.d.). Retrieved on April 13, 2020 from https://www.bjs.gov/index.cfm?ty=tp&tid=2.

Burkhardt, B.C. (2017, March 20). Private prisons, explained. *The Conversation*. Retrieved on April 28, 2020 from https://theconversation.com/private-prisons-explained-73038.

Burt, M. (1981). *Measuring prison results: Ways to monitor and evaluate corrections performance.* Washington, DC: National Institute of Justice.

Cali, J. (2013, March 12). Frequent reference question: How many federal laws are there? *The Library of Congress, Blogs, Law Library.* Retrieved on July 4, 2020 from https://blogs.loc.gov/law/2013/03/frequent-reference-question-how-many-federal-laws-are-there/.

Caulkins, J.P. (1993). Zero-tolerance policies: Do they inhibit or stimulate illicit drug consumption? *Management Science, 29*(4), 458–476.

Caulkins, J.P., Cohen, J., Gorr, W., & Wei, J. (1996). Predicting criminal recidivism: A comparison of neural network models with statistical methods. *Journal of Criminal Justice, 24*(3), 227–240.

Celona, L., Feuerherd, B., & Weissmann, R. (2020, January 11). Controversial bail reform springs serial robbery suspect-who then pulls off fifth heist. *New York Post.* Retrieved on March 30, 2020 from https://nypost.com/2020/01/11/serial-robber-released-with-no-bail-then-immediately-robs-another-bank/.

Chaiken, J., Crabill, T., Holliday, L., Jaquette, D., Lawless, M., & Quade, E. (1976). *Criminal justice models: An overview.* Washington, DC: National Institute of Law Enforcement and Criminal Justice, Law Enforcement Assistance Administration.

Chapman, A. (n.d.). How long do felony cases take? Retrieved on March 26, 2020 from https://www.amychapmanlaw.com/practice-areas/criminal-defense/how-long-do-felony-cases-take/.

Clemen, R.T., & Reilly, T. (2013). *Making hard decisions with decision tools.* 3rd ed. Mason, OH: South-Western Cengage Learning.

Community policing grants: COPS grants were a modest contributor to declines in crime in the 1990s. (2005, October). US Government Accountability Office Report. Retrieved on July 16, 2020 from https://www.gao.gov/new.items/d06104.pdf.

Cooprider, K. (2009, June). Pretrial risk assessment and case classification: A case study control. *Federal Probation, 73*(1). June 2009 Newsletter. Retrieved on March 30, 2020 from https://www.uscourts.gov/federal-probation-journal/2009/06/pretrial-risk-assessment-and-case-classification-case-study.

Cost of Crime Calculator. (n.d.). RAND, social and economic well-being. Retrieved on August 3, 2020 from https://www.rand.org/well-being/justice-policy/centers/quality-policing/cost-of-crime.html.

Creed, W. (2019, March 10). First woman released under the first step act. *Cape Charles Mirror.* Retrieved on June 2, 2020 from http://www.capecharlesmirror.com/news/first-woman-released-under-the-first-step-act/.

Criminal Justice Lab, NYU Law. (n.d.). Retrieved on April 21, 2020 from https://www.law.nyu.edu/centers/cjl.

Criminal Justice Reform. (n.d.). The first step act of 2018. Retrieved on April 22, 2020 from https://www.whitehouse.gov/wp-content/uploads/2020/02/FY21-Fact-Sheet-Criminal-Justice-Reform.pdf.

Criminal Justice System. (n.d.). Lexico. Oxford English and Spanish dictionary. Retrieved on August 8, 2020 from https://www.lexico.com/definition/criminal_justice_system.

Data Collection: Census of Problem Solving Courts. (2012). Bureau of Justice Statistics. Retrieved on August 8, 2020 from https://www.bjs.gov/index.cfm?ty=dcdetail&iid=448.

Data Collection: National Crime Victimization Survey (NCVS). (2018). Bureau of Justice Statistics. Retrieved on August 8, 2020 from https://www.bjs.gov/index.cfm?ty=dcdetail&iid=245.

Demko, P. (2016, August 7). How Pence's slow walk on needle exchange helped propel Indiana's health crisis. *Politico*. Retrieved on June 21, 2020 from https://www.politico.com/story/2016/08/under-pences-leadership-response-to-heroin-epidemic-criticized-as-ineffective-226759.

Department of Justice. (2019a, July 19). *Department of Justice announces the release of 3,100 inmates under first step act, publishes risk and needs assessment system.* Retrieved on May 30, 2020 from https://www.justice.gov/opa/pr/department-justice-announces-release-3100-inmates-under-first-step-act-publishes-risk-and.

Department of Justice. (2019b, July 19). *The first step act of 2018: Risk and needs assessment system.* Retrieved on May 30, 2020 from https://nij.ojp.gov/sites/g/files/xyckuh171/files/media/document/the-first-step-act-of-2018-risk-and-needs-assessment-system_1.pdf.

Des Jarlais, D.C., Perlis, T., Arasteh, K., Torian, L.V., Beatrice, S., Milliken, J., Mildvan, D., Yancovitz, S., & Friedman, S.R. (2005). HIV incidence among injection drug users in New York City, 1990 to 2002: Use of serologic test algorithm to assess expansion of HIV prevention services. *American Journal of Public Health, 95*(8), 1439–1444.

deVuono-powell, S., Schweidler, C., Walters, A., & Zohrabi, A. (2015, September). *Who pays? The true cost of incarceration on families.* Oakland, CA: Ella Baker Center, Forward Together, Research Action Design. Retrieved on March 28, 2020 from https://ellabakercenter.org/sites/default/files/downloads/who-pays.pdf.

Drusch, A. (2018a, April 13). Washington looks to Texas on federal prison reform. *Ft. Worth Star Telegram*. Retrieved on May 30, 2020 from https://www.star-telegram.com/news/politics-government/article208733524.html.

Drusch, A. (2018b, December 14). Washington's prison reforms estimated to cost $346M over 10 years. *Ft. Worth Star Telegram.* Retrieved on May 29, 2020 from https://www.star-telegram.com/latest-news/article223049950.html.

Durose, M.R., Cooper, A.D., & Snyder, H.N. (2014). Recidivism of prisoners released in 30 states in 2005: Patterns from 2005 to 2010. Bureau of Justice Statistics. Retrieved on June 30, 2020 from http://www.bjs.gov/content/pub/pdf/rprts05p0510.pdf.

Eden, C., & Ackermann, F. (2004). Cognitive mapping expert views for policy analysis in the public sector. *European Journal of Operational Research, 152,* 615–630.

Edwards, W. (1977). How to use multiattribute utility measurement for social decision making. *IEEE Transactions on Systems, Man, and Cybernetics, SMC, 7,* 326–340.

Edwards, W., & Barron, F.H. (1994). SMARTS and SMARTER: Improved simple methods for multiattribute utility measurement. *Organizational Behavior Human Decision Processes, 60*(3), 306–325.

Ehrlich, I. (1972). The deterrent effect of criminal law enforcement. *Journal of Legal Studies, 1,* 259–276.

Eisen, L.B. (2019, September 9). The 1994 crime bill and beyond: How federal funding shapes the criminal justice system. Brennan Center for Justice. Retrieved on July 13, 2020 from https://www.brennancenter.org/our-work/analysis-opinion/1994-crime-bill-and-beyond-how-federal-funding-shapes-criminal-justice.

El Sayed, S.A., Morris, R.G., DeShay, R.A., & Piquero, A.R. (2020, August). Comparing the rates of misconduct between private and public prisons in Texas. *Crime and Delinquency, 66* (9), 1217–1241.

Ellspermann, S.J., Evans, G.W., & Basadur, M. (2007). The impact of training on the formulation of ill-structured problems. *Omega, 35*(2), 221–236.

Evans, G.W. (2017). *Multiple criteria decision analysis for industrial engineering: Methodology and applications.* Boca Raton, FL: CRC Press, Taylor and Francis Group.

Fair Sentencing Act of 2010 (2010, August 3). Public law 111–220. Retrieved on June 4, 2020 from https://www.congress.gov/111/plaws/publ220/PLAW-111publ220.pdf.

Farley, R. (2016, April 12). Bill Clinton and the 1994 crime bill. *FactCheck.org.* Retrieved on July 23, 2020 from https://www.factcheck.org/2016/04/bill-clinton-and-the-1994-crime-bill/.

Fass, S.M., & Pi, C.-R. (2002). Getting tough on juvenile crime: An analysis of costs and benefits. *Journal of Research in Crime and Delinquency, 39*(4), 363–399.

FBI: UCR, 2017 Crime in the United States. (n.d.). Retrieved on August 3, 2020 from https://ucr.fbi.gov/crime-in-the-u.s/2017/crime-in-the-u.s.-2017/topic-pages/clearances.

Figueroa, T. (2018, September 30). When people on bail commit new crimes, they're often linked to drugs. *San Diego Union-Tribune*. Retrieved on March 30, 2020 from https://www.sandiegouniontribune.com/news/public-safety/sd-me-bail-crime-data-20180926-story.html.

FL S.B. 238. (2012). Privatization of correctional facilities. Retrieved on May 16, 2020 from https://flsenate.gov/Session/Bill/2012/2038/BillText/e1/HTML.

Gaes, G.G. (2019). Current status of prison privatization research on American prisons and jails. *Criminology and Public Policy, 18,* 269–293.

GAO Report. (2017). Costs of crime: Experts report challenges estimating costs and suggest improvements to better inform policy decisions. GAO-17-732. Retrieved May 19, 2018, from https://www.gao.gov/products/GAO-17-732

Geng, L. (2018, June 29). Do more than 450,000 Americans sit in jail because they are too poor to pay bail? *Politifact: The Poynter Institute*. Retrieved on March 5, 2020 from https://www.politifact.com/factchecks/2018/jun/29/ted-lieu/do-more-450000-americans-sit-jail-because-they-are/.

Gest, T. (2019, September 13). 1994 crime bill's effect on mass incarceration was limited: Study. *The Crime Report: Your Criminal Justice Network*. Retrieved on July 21, 2020 from https://thecrimereport.org/2019/09/13/crime-bills-effect-on-mass-incarceration-limited-argues-new-study/.

Goldberg, N. (2020, February 24). Accused serial bank robber held without bail on federal charges because he has no family with money. *New York Daily News*. Retrieved on March 1, 2020 from https://www.nydailynews.com/new-york/nyc-crime/ny-bank-robbery-woodberry-bail-reform-20200224-fsxv6va62r-deboxolkwb6jwygm-story.html.

Gornoski, D. (2019, April 1). First black woman released under prison reform thanks God and President Trump. *Townhall Finance*. Retrieved on June 4, 2020 from https://finance.townhall.com/columnists/davidgornoski/2019/04/01/first-black-woman-released-under-prison-reform-thanks-god-and-president-trump-n2544070.

Gorski, C. (2012, March 23). The mathematics of jury size. *Inside Science*. Retrieved on July 15, 2020 from https://www.insidescience.org/news/mathematics-jury-size.

Harris, C.M., & Thlagarajan, T.R. (1975). Queueing models of community correctional centers in the District of Columbia. *Management Science, 22*(2), 167–171.

Haynes, C. (2018, August 30). The first step act: A pros and cons list. Equal justice under law. Retrieved on May 29, 2020 from https://equaljusticeunderlaw.org/thejusticereport/2018/8/21/the-first-step-act-a-pros-and-cons-list.

Hedger, L.A. (2017, October 26). 2nd Indiana county ends needle exchange, with one official citing moral concerns. *Indianapolis Star*. Retrieved on June 21, 2020 from https://www.indystar.com/story/news/2017/10/23/2nd-indiana-county-ends-needle-exchange-one-official-citing-moral-concerns/787740001/.

Hennessy-Fiske, M. (2015, December 21). Texas grand jury finds no cause for indictment in Sandra Bland case. *Los Angeles Times*. Retrieved on March 1, 2020 from https://www.latimes.com/nation/nationnow/la-na-nn-sandra-bland-grand-jury-20151221-story.html.

Henrichson, C., Rinaldi, J., & Delaney, R. (2015, May). The price of jails: Measuring the taxpayer cost of local incarceration. Report from the Center on Sentencing and Corrections, Vera Institute of Justice. Retrieved on March 28, 2020 from https://www.federalregister.gov/documents/2018/04/30/2018-09062/annual-determination-of-average-cost-of-incarceration.

Highest to Lowest Prison Population Total. (n.d.). World Prison Brief. Retrieved on August 9, 2020 from https://prisonstudies.org/highest-to-lowest/prison-population-total.

Hing, J. (2016). The democratic presidential candidates would end private prison contracts. *The Nation*. Retrieved on May 1, 2020 from https://www.thenation.com/article/archive/the-democratic-presidential-candidates-would-end-private-prison-contracts/.

Hinton, J. (2019, December 6). Neighborhood group wants stricter guidelines for needle exchange program. *News 13*, WLOS, Asheville, NC. Retrieved on June 23, 2020 from https://wlos.com/news/local/neighborhood-group-wants-stricter-guidelines-for-needle-exchange-programs.

Ho, V. (2018, December 19). Criminal justice reform bill passed by Senate in rare bipartisan victory. *The Guardian*. Retrieved on May 29, 2020 from https://www.theguardian.com/us-news/2018/dec/18/first-step-act-criminal-justice-reform-passes-senate.

Hopkins, B., Bains, C., & Doyle, C. (2018). Principles of pretrial release: Reforming bail without repeating its harms. *Journal of Criminal Law & Criminology, 108*(4), 679–700. Retrieved on March 30, 2020 from https://scholarlycommons.law.northwestern.edu/jclc/vol108/iss4/2/.

How much are attorney fees? (n.d.). Thervo. Retrieved on July 1, 2020 from https://thervo.com/costs/attorney-fees.

H.R. 5484 (99th): Anti-Drug Abuse Act of 1986. (n.d.). Govtrack. Retrieved on April 30, 2020 from https://www.govtrack.us/congress/bills/99/hr5484.

Josephson, A. (2018, June 25). The economics of bail. Published by smartasset.com. Retrieved on January 15, 2020 from https://smartasset.com/personal-loans/the-economics-of-bail.

Kaeble, D., Maruschak, L.M., & Bonczar, T.P. (2015, November). *Probation and parole in the United States, 2014*. U.S. Department of Justice Office of Justice Programs. Bureau of Justice Statistics (NCJ 249057). Retrieved on June 30, 2020 from http://www.bjs.gov/content/pub/pdf/ppus14.pdf.

Kaplan, E.H., & O'Keefe, E. (1993, January). Let the needles do the talking! Evaluating the New Haven needle exchange. *Interfaces, 23*(1), 7–26.

Keeney, R.L. (1992). *Value focused thinking: A path to creative decision making*. Cambridge, MA: Harvard University Press.

Keeney, R.L. (2012). Value-focused brainstorming. *Decision Analysis, 9*(4), 303–313.

Keeney, R.L., & Gregory, R.S. (2005). Selecting attributes to measure the achievement of objectives. *Operations Research, 53*(1), 1–11.

Keeney, R.L., & Raiffa, H. (1993). *Decisions with multiple objectives: Preferences and value tradeoffs*. New York: Cambridge University Press.

Kelton, W.D., Sadowski, R.P., & Zupick, N.B. (2015). *Simulation with arena*. 5th ed. New York: McGraw-Hill Education.

Kepner, C.H., & Tregoe, B.B. (1981). *The new rational manager*. Princeton, NJ: Princeton Research Press.

Lanza-Kaduce, L., Parker, K.F., & Thomas, C.W. (1999, January 1). A comparative recidivism analysis of releasees from private and public prisons. *Crime & Delinquency, 45*(1), 28–47.

Larson, R.C., Cahn, M.F., & Shell, M.C. (1993). Improving the New York City arrest-to-arraignment System. *Interfaces, 23*(1), 76–96.

Law, A.M. (2007). *Simulation modeling and analysis*. New York: McGraw-Hill.

Law Enforcement Officers Killed and Assaulted (2019). Tables (2020, June 18). Bureau of Justice Statistics. Retrieved on August 8, 2020 from https://www.bjs.gov/index.cfm?ty=pbdetail&iid=6906.

Layne, R. (2019, June 27). Private prisons get slammed as democrats debate. *CBS News*. Retrieved on May 1, 2020 from https://www.cbsnews.com/news/private-prisons-get-slammed-ahead-of-democratic-debates/.

Linstone, H., & Turoff, M. (1975). *The Delphi method: Techniques and applications*. Boston, MA: Addison-Wesley Publishing Company.

Logan, C.H. (1993, October). Criminal justice performance measures for prisons. Pages 19 to 60 in U.S. Department of Justice Report, DiIulio, Jr., J.J. et al. *Performance measures for the criminal justice system*, Retrieved on May 18, 2020 from https://www.bjs.gov/content/pub/pdf/pmcjs.pdf.

Lombardo, C. (2014, November 30). Pros and cons of privatization of prisons. *Vision Launch Media*. Retrieved on April 27, 2020 from https://visionlaunch.com/pros-and-cons-of-privatization-of-prisons/.

Lopez, G. (2018, September 24). There's a nearly 40% chance you'll get away with murder in America. *Vox*. Retrieved on August 3, 2020 from https://www.vox.com/2018/9/24/17896034/murder-crime-clearance-fbi-report.

Lowenkamp, C.T., VanNostrand, M., & Holsinger, A. (2013, November). Investigating the impact of pretrial detention on sentencing outcomes. *Laura and John Arnold Foundation*. Retrieved on March 6, 2020 from https://university.pretrial.org/HigherLogic/System/DownloadDocumentFile.ashx?DocumentFileKey=172dd7bf-96cf-aa8d-75d0-399b1a9b17e3&forceDialog=0.

Lundahl, B.W., Kunz, C., Brownell, C., Harris, N., & Van Fleet, R. (2009). Prison privatization: A meta-analysis of cost and quality of confinement indicators. *Research on Social Work Practice, 19*(4), 383–394.

Lussenhop, J. (2016, April 18). Clinton crime bill: Why is it so controversial? *BBC News Magazine*. Retrieved on July 16, 2020 from https://www.bbc.com/news/world-us-canada-36020717.

Maltz, M.D. (1994). Operations research in studying crime and justice: Its history and accomplishments. In S.M. Pollock, M.H. Rothkopf, & A. Barnett (Eds.), *Operations research and the public sector*. Amsterdam: North-Holland.

Manning, M., Wong, T.W.G., & Vorsina, M. (2016). *Manning cost-benefit tool*. Canberra: ANU Centre for Social Research and Methods, The Australian National University. Retrieved on June 13, 2020 from research/projects/manning-cost-benefit-tool.

Manning, M., Wong, G.T.W., Graham, T., Ranbaduge, T., Christen, P., Taylor, K., Wortley, R., Makkai, T., & Skorich, P. (2018). Towards a smart cost-benefit tool: Using machine learning to predict the costs of criminal justice policy interventions. *Crime Science, 7*(12). Retrieved on June 12, 2020 from https://crimesciencejournal.biomedcentral.com/articles/10.1186/s40163-018-0086-4.

Marvell, T.B., & Moody, C.E. (1996). Specification problems, police levels, and crime rates. *Criminology, 34*, 609–646.

Mello, S. (2018, February 25). More cops, less crime. Retrieved on June 10, 2020 from https://www.princeton.edu/~smello/papers/cops.pdf.

Merlone, U., Manassero, M., & Raffaello, C. (2016). The lingering effects of past crimes over future criminal careers. In *Proceedings of the 2016 Winter Simulation Conference*, 3532–3542, December 11–14, 2016, Arlington, VA

Merriam Webster Dictionary. (n.d.). Definition of law. Retrieved on August 26, 2020 from https://www.merriam-webster.com/dictionary/law.

Miller, J. (1989, April 23). Criminology. *The Washington Post*. Retrieved on July 27, 2020 from https://www.washingtonpost.com/archive/opinions/1989/04/23/criminology/3e8fb430-9195-4f07-b7e2-c97a970c96fe/.

Monk, J. (2019). Freed SC inmate thanks Trump. 'I'm determined not to let my past define my future'. *The State*. Retrieved on June 3, 2020 from https://www.thestate.com/news/politics-government/election/article236621348.html.

Moos, R. (1987). Correctional institutions environment scale. *Mind Garden*. Retrieved on May 17, 2020 from https://www..mindgarden.com/89-correctional-institutions-environment-scale.

Mumford, M., Schanzenbach, D.W., & Nunn, R. (2016, October 20). The economics of private prisons. *The Hamilton Project*. Retrieved on March 20, 2020 from https://www.hamiltonproject.org/papers/the_economics_of_private_prisons.

Nadler, G., & Hibino, S. (1990). *Breakthrough thinking*. Rocklin, CA: Prima Publishing.

Nagel, S.S. (1981). Management science and jury size. *Interfaces*, *11*(3), 34–39.

Operations Research and Analytics. (n.d.). INFORMS web site. Retrieved on August 21, 2020 from https://www.informs.org/Explore/Operations-Research-Analytics.

OptQuest: The World's Leading Simulation Optimization Engine. (n.d.). OptTek. Retrieved on July 9, 2020 from https://www.opttek.com/products/optquest/.

Oudekerk, B. (2019, March 29). Hate crime statistics. Bureau of Justice Statistic Presentation. Retrieved on August 8, 2020 from https://www.bjs.gov/index.cfm?ty=pbdetail&iid=6906.

Pauley, M. (2016, July/August). A brief history of America's private prison industry. *Mother Jones*. Retrieved on April 30, 2020 from https://www.motherjones.com/politics/2016/06/history-of-americas-private-prison-industry-timeline/.

Perrone, D., & Pratt, T.C. (2003). Comparing the quality of confinement and cost-effectiveness of public versus private prisons: What we know, why we do not know more, and where to go from here. *The Prison Journal*, 83(3), 301–322.

Pfaff, J.F. (2016). The complicated economics of prison reform, *Michigan Law Review, 114*, 951–981.

Pratt, T.C. (2019). Cost-benefit analysis and privatized corrections. *Criminology & Public Policy, 18*(2), 447–456.

Pratt, T.C., & Maahs, J. (1999). Are private prisons more cost-effective than public prisons? A meta-analysis of evaluation research studies. *Crime & Delinquency, 45*, 358–371.

Private Jails in the United States. (2017, July 28). FindLaw. Retrieved on April 30, 2020 from https://civilrights.findlaw.com/other-constitutional-rights/private-jails-in-the-united-states.html.

Private Prisons in the United States. (2019, October 24). The Sentencing Project. Retrieved on April 30, 2020 from https://www.sentencingproject.org/publications/private-prisons-united-states/.

Rahman, I. (2019, July). New York, New York: Highlights of the 2019 bail reform law. Vera Institute of Justice Report. Retrieved on January 28, 2020 from https://www.vera.org/downloads/publications/new-york-new-york-2019-bail-reform-law-highlights.pdf.

Rennison, C.M., & Dodge, M. (2018). *Introduction to criminal justice: Systems, diversity, and change.* 2nd ed. Thousand Oaks, CA: Sage Publications.

Rhodes, W., Gaes, G.G., Kling, R., & Cutler, C. (2018, August 8). Relationship between prison length of stay and recidivism: A study using regression discontinuity and instrumental variables with multiple break points. *Criminology & Public Policy, 17*(3), 731–769.

Rosenfeld, R. (2019, September). The 1994 crime bill, legacy and lessons: Overview and reflections. Council on Criminal Justice. Retrieved on July 21, 2020 from https://cdn.ymaws.com/counciloncj.org/resource/resmgr/crime_bill/overview:and_reflections.pdf.

Sabol, W.J., & Johnson, T.L. (2019, September). The 1994 crime bill, legacy and lessons. Part 1: Impacts on prison population. Council on Criminal Justice. Retrieved on July 21, 2020 from https://cdn.ymaws.com/counciloncj.org/resource/resmgr/crime_bill/part_one_-_prison_population.pdf.

Schwartzbach, M. (n.d.). How long do criminal trials take? *Nolo.* Retrieved on March 26, 2020 from https://www.nolo.com/legal-encyclopedia/how-criminal-cases-take.html.

Segal, G.F., & Moore, A.T. (2002, January). Weighing the watchmen: Evaluating the costs and benefits of outsourcing correctional services. Part II: Reviewing the literature on cost and quality comparisons (Reason Public Policy Institute Study No. 290).

Simon, H.A. (1960). *The new science of management.* New York: Harper and Row.

Simon, J. (2018, October 14). Adjusting the scales: The elimination of cash bail. *The University of Cincinnati Law Review*. Retrieved on January 20, 2020 from https://uclawreview.org/2018/10/14/adjusting-the-scales-the-elimination-of-cash-bail/.

Sipes, Jr. L.A. (2018, June). Offender recidivism and reentry in the United States. *Crime in America.net*. Retrieved on July 18, 2020 from https://www.crimeinamerica.net/offender-recidivism-and-reentry-in-the unitedstates/#:~:text=The%20most%20common%20understanding%20of%20recidivism%20is%20based,three-quarters%20%2877%20percent%29%20were%20arrested%20within%20five%20years.

Spivak, A.L., & Sharp, S.F. (2008). Inmate recidivism as a measure of private prison performance. *Crime & Delinquency, 54*(3), 482–508.

Stanglin, D. (2020, July 3). Fact Check: 1994 Crime Bill Did Not Bring Mass Incarceration of Black Americans. *USA Today*. Retrieved on July 13, 2020 from https://www.usatoday.com/story/news/fact-check/2020/07/03/fact-check-1994-crime-bill-didnt-bring-mass-incarceration-black-people/3250210001/.

Statistical Briefing Book. (n.d.). Office of Juvenile Justice and Delinquency Prevention. Retrieved on July 6, 2020 from https://www.ojjdp.gov/ojstatbb/court/qa06201.asp?qaDate=2018.

Stringer, S.M. (2018, January 17). The public cost of private bail: A proposal to ban bail bonds in NYC. *Report of the New York City Comptroller*. Retrieved on January 16, 2020 from https://comptroller.nyc.gov/reports/the-public-cost-of-private-bail-a-proposal-to-ban-bail-bonds-in-nyc/.

Sullivan, C.M., & O'Keeffe, C.P. (2016, July 25). Does more policing lead to less crime-or just more racial resentment? *The Washington Post*. Retrieved on July 17, 2020 from https://www.washingtonpost.com/news/monkey-cage/wp/2016/07/25/does-more-policing-lead-to-less-crime-or-just-more-racial-resentment/.

Summers, C., & Willis, T. (2010, October 18). Pretrial risk assessment. Report from the Bureau of Justice Assistance, US Bureau of Justice. Retrieved on February 3, 2020 from https://bja.ojp.gov/sites/g/files/xyckuh186/files/media/document/PretrialRiskAssessmentResearchSummary.pdf.

Syringe Service Programs (SSPs) FAQs. (2019, May 23). Centers for Disease Control and Prevention. Retrieved on June 23, 2020 from https://www.cdc.gov/ssp/syringe-services-programs-faq.html.

The Bail Reform Act of 1984. (n.d.). US legal. Retrieved on August 9, 2020 from https://bail.uslegal.com/bail-and-bail-bond-agents/the-bail-reform-act-of-1984/#:~:text=The%20Bail%20Reform%20Act%20of%201984.%20The%20Bail,resulted%20in%20defendants%20who%20committed%20further%20violent%20crimes.

The First Step Act of 2018. (2019). Retrieved on May 29, 2020 from https://www.whitehouse.gov/wp-content/uploads/2020/02/FY21-Fact-Sheet-Criminal-Justice-Reform.pdf.

The Nation's Two Measures of Homicide. (2014, July). US Department of Justice. *NCJ 247060*. Retrieved on August 8, 2020 from https://www.bjs.gov/content/pub/pdf/ntmh.pdf.

Top 10 reasons for crime (2019, October 8). *NetNewsLedger*. Retrieved on July 17, 2020 from http://www.netnewsledger.com/2019/10/08/top-10-reasons-for-crime/.

Tracy, T. (2020, January 11). Man robbed and robbed again; then he's arrested, freed on bail under new NY laws, and robs another bank in Brooklyn, cops say. *New York Daily News*. Retrieved on March 1, 2020 from https://www.nydailynews.com/new-york/nyc-crime/ny-released-bank-robber-released-without-bail-wanted-for-another-crime-20200111-4l4ow4kxgra23kjadl4rjsnkiq-story.html.

United States population and rate of crime per 100,000 people 1960–2018. (n.d.). Retrieved on July 29, 2020 from http://www.disaster-center.com/crime/uscrime.htm.

VanNostrand, M., & Keebler, G. (2009). Pretrial risk assessment in the Federal Court. *Federal Probation, 72*(2). Retrieved on March 30, 2020 from https://www.uscourts.gov/federal-probation-journal/2009/09/pretrial-risk-assessment-federal-court.

Violent Crime Control and Law Enforcement Act of 1994. (1994). U.S. Department of Justice. Retrieved on April 28, 2020 from https://www.ncjrs.gov/txtfiles/billfs.txt.

Volokh, A. (2002). A tale of two systems; cost, quality, and accountability in private prisons. *Harvard Law Review, 115*, 1868–1890.

Volokh, A. (2013). Prison accountability and performance measures. *Emory Law Journal, 63*, 339–416.

Von Winterfeldt, D., & Edwards, W. (1986). *Decision analysis and behavioral research*. London, UK: Cambridge University Press.

Wakefield, S. (2018, August 8). Sentence length and recidivism: Evidence and the challenges of criminal justice reform in the carceral state. *Criminology and Public Policy, 17*(3), 771–777.

Wallin, P. (n.d.). How long do you have to wait before going to court after you are arrested? *Walli and Klarich: A Law Corporation*. Retrieved on March 5, 2020 from https://www.wklaw.com/wait-before-going-to-court/.

Washington State Institute for Public Policy (n.d.). Benefit cost results. Retrieved on July 1, 2020 from http://www.wsipp.wa.gov/BenefitCost/Pdf/2/WSIPP_BenefitCost_Adult-Criminal-Justice.

World Justice Project Rule of Law Index 2020 (2020). World Justice Project. Retrieved on August 9, 2020 from https://worldjusticeproject.org/sites/default/files/documents/WJP-ROLI-2020-Online_0.pdf.

Worrall, J.L., & Kovandzic, T.V. (2007, February). Cops grants and crime revisited. *Criminology, 45*(1), 159–190.

Yablon, M. (1991). Modeling prison populations. *European Journal of Operational Research, 52,* 259–266.

Zarkin, G.A., Cowell, A J., Hicks, K.A., Mills, M.J., Belenko, S., Dunlap, L.J., & Keyes, V. (2015). Lifetime benefits and costs of diverting substance-abusing offenders from state prison. *Crime and Delinquency, 61*(6), 829–850.

Zeng, Z. (2018, February). Jail inmates in 2018. Report NCJ 251210 from the Bureau of Justice Statistics, U.S. Department of Justice. Retrieved on March 5, 2020 from https://www.bjs.gov/index.cfm?ty=pbdetail&iid=6826.

Zhao, J., Scheider, M.C., & Thurman, Q.C. (2002). Funding community policing to reduce crime: Have COPS grants made a difference? *Criminology and Public Policy, 2,* 7–32.

INDEX

Printed in the United States
by Baker & Taylor Publisher Services